D1824718

Hypercomplex Universe: Unified SuperStandard Theories UST, QUeST, MOST

Stephen Blaha Ph. D.
Blaha Research

Scalar Dimension Fields
Dimension Fields May Be Higgs Bosons
Fermion-Dimension SuperSymmetry
Fermionic Operators
Underlying Framework of UST, QUeST, and MOST
One Dimension–One Fermion Proto-Universe
Any Fermion Transformable to Any Other Fermion
Beyond Space-Time in QUeST

Pingree-Hill Publishing
MMXX

Copyright © 2020 by Stephen Blaha. All Rights Reserved.

This document is protected under copyright laws and international copyright conventions. No part of this book may be reproduced, stored in a retrieval system, or transmitted by any means in any form, electronic, mechanical, photocopying, recording, or as a rewritten passage(s), or otherwise, without the express prior written permission of Blaha Research. For additional information send an email to the author at sblaha777@yahoo.com or call 603-289-5435.

ISBN: 978-1-7345834-7-2

This document is provided "as is" without a warranty of any kind, either implied or expressed, including, but not limited to, implied warranties of fitness for a particular purpose, merchantability, or non-infringement. This document may contain typographic errors, technical inaccuracies, and may not describe recent developments. This book is printed on acid free paper.

Rev. 00/00/01   July 21, 2020

To Margaret

Some Other Books by Stephen Blaha

All the Megaverse! Starships Exploring the Endless Universes of the Cosmos using the Baryonic Force (Blaha Research, Auburn, NH, 2014)

SuperCivilizations: Civilizations as Superorganisms (McMann-Fisher Publishing, Auburn, NH, 2010)

All the Universe! Faster Than Light Tachyon Quark Starships & Particle Accelerators with the LHC as a Prototype Starship Drive Scientific Edition (Pingree-Hill Publishing, Auburn, NH, 2011).

Unification of God Theory and Unified SuperStandard Model THIRD EDITION (Pingree Hill Publishing, Auburn, NH, 2018).

The Exact QED Calculation of the Fine Structure Constant Implies ALL 4D Universes have the Same Physics/Life Prospects (Pingree Hill Publishing, Auburn, NH, 2019).

Unified SuperStandard Theory and the SuperUniverse Model: The Foundation of Science (Pingree Hill Publishing, Auburn, NH, 2018).

Quaternion Unified SuperStandard Theory (The QUeST) and Megaverse Octonion SuperStandard Theory (MOST) (Pingree Hill Publishing, Auburn, NH, 2020).

Unified SuperStandard Theories for Quaternion Universes & The Octonion Megaverse (Pingree Hill Publishing, Auburn, NH, 2020).

The Essence of Eternity: Quaternion & Octonion SuperStandard Theories (Pingree Hill Publishing, Auburn, NH, 2020).

Available on Amazon.com, bn.com Amazon.co.uk and other international web sites as well as at better bookstores (through Ingram Distributors).

CONTENTS

FIGURES and TABLES

ESSENCE OF ETERNITY

known fermions including an additional, as yet not found, 4th generation shown. The lines on the left side (only shown for one layer) display the Generation mixing within each layer's species. The Generation mixing applies within each layer using a separate Generation group for each layer. The lines on the right side show Layer group mixing with the mixing amongst all four layers for each of the four generations individually. There are four Layer groups. The Dark groups mixing between normal and Dark fermions are shown in the center as horizontal lines. For each generation and each layer SU(2) mixes between an e-type fermion and a neutrino-type fermion. It also mixes between an up-quark-type fermion and a down-quark-type fermion. SU(3) mixes among each up-quark triplet and down-quark triplet separately. Complex Lorentz group transformations map among all four fermions: Dirac $\leftrightarrow$ tachyon $\leftrightarrow$ up-quark $\leftrightarrow$ down-

INTRODUCTION

This book describes the detailed relation of quaternion QUeST to the Unified SuperStandard Theory of Blaha (2020c) and earlier books. The Unified SuperStandard Theory (UST) in our 3 + 1 dimension space-time was derived from Complex General Relativity and Quantum Field Theory suitably extended. We have shown that 32 complex quaternion dimension QUeST has an identical pattern of Internal Symmetries as UST. *UST is derivable from QUeST.*

This remarkable coincidence led us to explore Unified SuperStandard Theories in greater detail in this book, and in the two *Essence of Eternity* books which are included here for the convenience of readers.

The QUeST (and UST) universe relate particle dynamics directly to complex quaternion space in a manner analogous to the corresponding relation between gravitational dynamics and General Relativity. A similar relationship exists for MOST and the Megaverse. This is evident in all the 2020 quaternion books by the author.

This book continues the examination of particle-dimension duality, and goes beyond earlier books with a 256 dimension space by developing a possible, four-dimension basis, and a yet more basic one dimension – one fermion theory.

It also develops a fermion-dimension SuperSymmetry with fermionic operators that suggestively lead to a possible basis for Higgs bosons.

The book also explores Megaverse fermion-dimension duality.

Lastly, we discuss the possibilities of probing internal symmetry dimensions of QUeST in our universe, and of probing exits from our universe into the larger space of the Megaverse.

1. A Deeper Basis for 32 Dimension Complex Quaternion QUeST

Into what is the Universe Resolved?

Having successfully based the Unified SuperStandard Theory (UST) on the Quaternion Unified SuperStandard Theory (QUeST) with a remarkable match between the internal symmetries and space-time symmetry of both theories, we now encounter the question: *Is there a yet deeper basis for QUeST?* It appears there is.

QUeST was formulated in 32 dimension complex quaternion space with a total number of 256 dimensions. The number of fundamental fermions was found to number 256. Fig. 1.1 lists the fermion periodic table in UST and QUeST including the currently known fermions. It also shows the interactions between the fermions. Notice all fermions are connected in the sense that interactions exist to transform any fermion into any other fermion. *The interconnections of the fermions suggest that there is a commonality between them, which the deeper basis proposed in this chapter supports.*

Fig. 1.2 shows the correspondence between fermions and dimensions which clearly extends down to the level of generations. Each fermion generation with 8 fermions has a corresponding complex quaternion with 8 dimensions—a clear-cut rationale for fermion generations.

This chapter provides a derivation of a new basis for QUeST and UST. We will call it the q-Basis.[1]

1.1 The Nature of Dimensions

For 2500 years, since Euclid and earlier, dimensions have been simply counted as integers in the definition of coordinate systems. The nature of a dimension seems to consist of two parts: an "identifier" plus a one-dimensional set of coordinates. A finite-

---

[1] Similarly we develop a new deeper basis for MOST in the Megaverse.

dimensional space of dimension n consists of many dimensions and an n-tuple of coordinates.

Fig. 1.1 compares a dimension and a fermion field. A field is defined as the functional inner product[2] of a particle functional and a coordinate wave as described in Blaha (2020c) and earlier books.[3] Fig. 1.3 displays the (fundamental) Fermion Periodic Table with the interactions that transform between fermions in UST and QUeST further solidifying the relation of dimensions to fermions. Fig. 1.2 shows a row by row match between dimensions and fermions using a dot (pebble) symbol to represent dimensions and fermions. The pebble symbols were introduced around 500 BCE by the Pythagoreans. (See Blaha (2020e) for more detail.)

Dimension	**Particle Field**
Identifier	Functional
One dimensional set of coordinates	A set of Fourier coordinates

Figure 1.1. Comparison of a dimension and a quantum field defined with a functional.

A comparison suggests it is reasonable to define a dimension, D, as a particle-like construct that is the inner product of a dimension functional, f, and a one-dimensional set of coordinates (x).

$$D = D(x) = (f, (x))$$ (1.1)

---

[2] See Riesz (
1955) where linear functionals and their inner products are defined.
[3] This definition of a quantum field supports relativistic quantum entanglement without paradoxes as we show in earlier books.

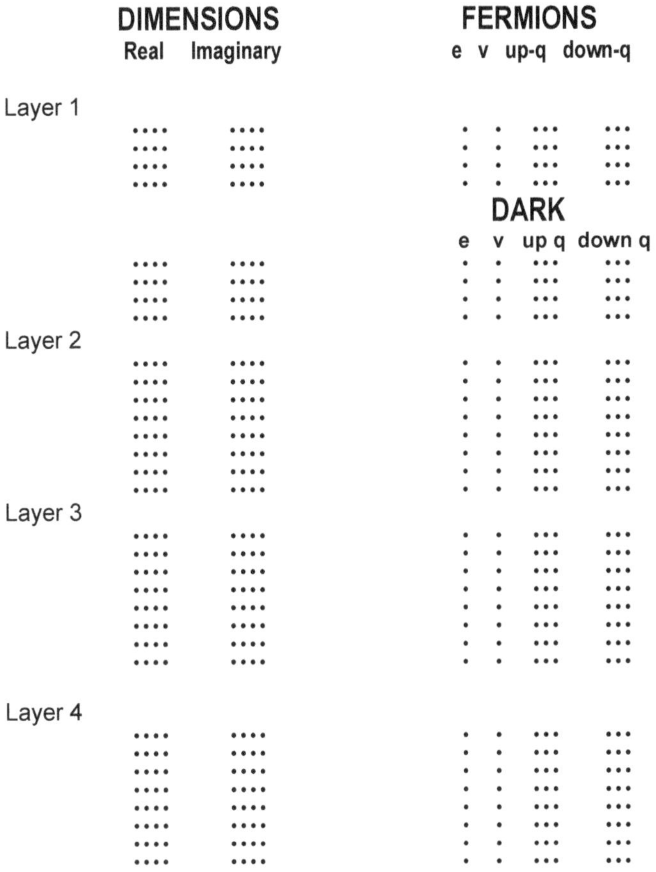

Figure 1.2. Fundamental fermions have a 1:1 correspondence with QUeST dimensions. Note the number of dimensions in each row is 8 – the number of dimensions in a complex quaternion. Correspondingly the number of fermions in each row is 8 – a suggestive similarity. Each layer has four normal fermion generations and four Dark fermion generations. Each dot (pebble) represents a dimension in the left part and a fermion in the right part.

The Fermion Periodic Table

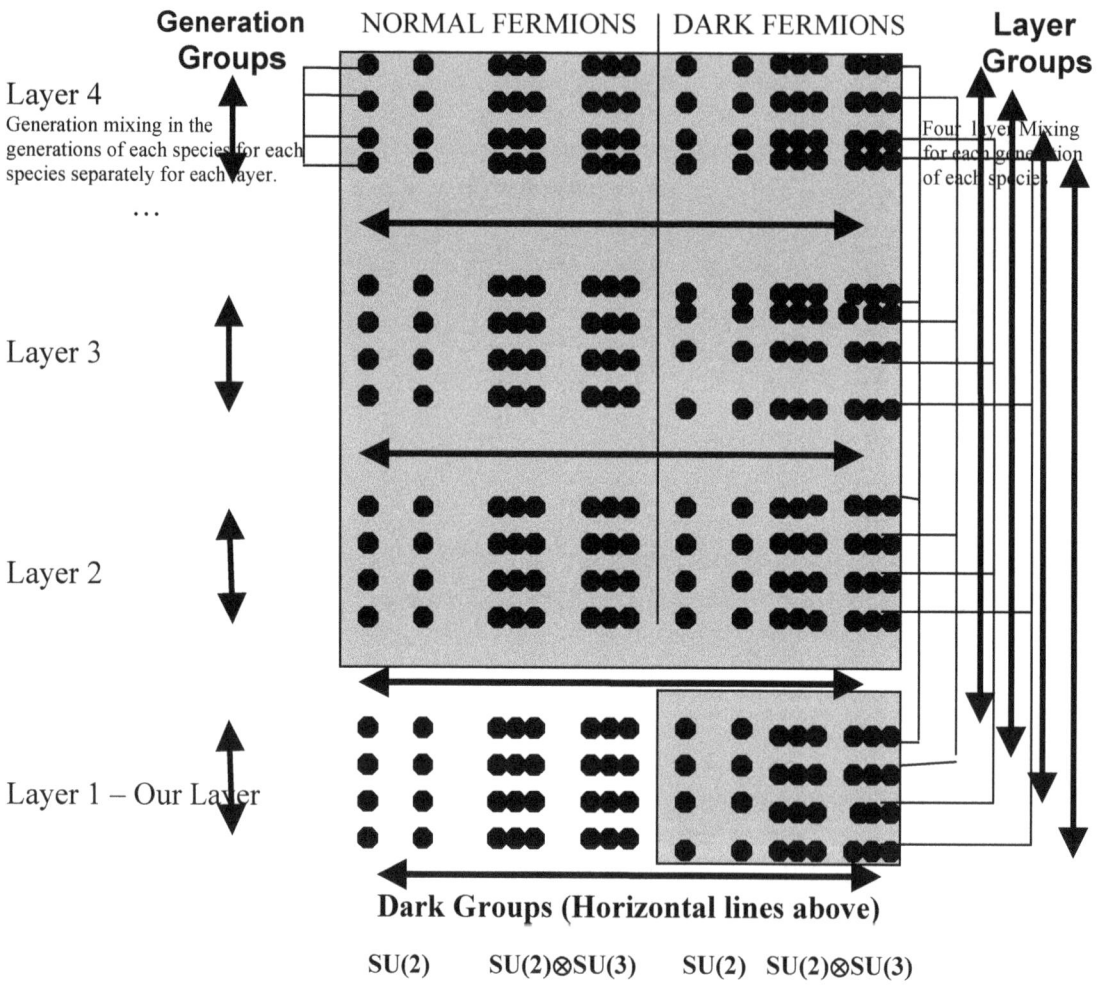

Generation Groups

NORMAL FERMIONS | DARK FERMIONS

Layer Groups

Layer 4
Generation mixing in the generations of each species for each species separately for each layer.

Four Layer Mixing for each generation of each species.

...

Layer 3

Layer 2

Layer 1 – Our Layer

Dark Groups (Horizontal lines above)

SU(2)   SU(2)⊗SU(3)   SU(2)   SU(2)⊗SU(3)

Complex Lorentz Group

Figure 1.3. Fermion particle spectrum and partial example of pattern of mass mixing of the Generation, Layer, and Dark grroups. Unshaded parts are the known fermions including an additional, as yet not found, 4th generation shown. The lines on the left side (only shown for one layer) display the Generation mixing within each layer's species. The Generation mixing applies within each layer using a separate Generation group for each layer. The lines on the right side show Layer group mixing with the mixing amongst all four layers for each of the four generations individually. There are four Layer groups. The Dark groups mixing between normal and Dark fermions are shown in the center as horizontal lines. For each generation and each layer SU(2) mixes between an e-type fermion and a neutrino-type fermion. It also mixes between an up-quark-type fermion and a down-quark-type fermion. SU(3) mixes among each up-quark triplet and down-quark triplet separately. Complex Lorentz group transformations map among all four fermions: Dirac ↔ tachyon ↔ up-quark ↔ down-quark. There are 256 fundamental fermions counting quarks as triplets.

A dimension then partakes of the features of a field. We call it a *dimension field*. We will denote dimension fields as $\varphi_i(x)$ where $i = 1, 2, \ldots, 256$.

   A further comparison of dimensions and the fermion spectrum in Fig. 1.2 suggests that we treat *dimension fields* as off shell spin ½ Dirac fields.

   We now define a 16×16 array[4] of dimensions, which places normal and Dark blocks of dimensions side by side. This is physically reasonable in view of Fig. 1.2 and 1.3. See Fig. 1.4, which corresponds to the fermion periodic table in Fig. 1.3.

   We further define the dimension array in Fig. 1.4

$$D_{ij} = D_{ij}(x_{ij}) = (f_{ij}, (x_{ij})) \qquad (1.2)$$

---

[4] We could have provisionally treated QUeST as based on a 16-dimension complex octonion space. This space has an interesting generation mechanism. If we define a complex octonion functional O_c that expands a complex octonion vector o_c so that each element of the vector is made into a complex octonion then the 16 by 16 array of Fig. 1.4 is generated by the simple expression:

$$O_c(o_c)$$

for i, j = 1, 2, …, 16. Then we create a corresponding 16×16 array of functionals

$$f(i, j) \tag{1.3}$$

using arguments instead of suffixes for typographic convenience.

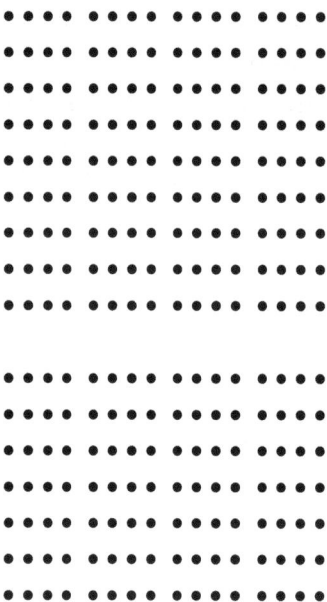

Figure 1.4. 16×16 array of dimensions.

Next we will create a deeper basis, q-basis, for QUeST by factoring the functional array. Our purpose is to reduce the number of dimensions (256), to a more "acceptable" number (16).

1.2 Derivation of a Deeper Basis: The q-Basis

We now establish a map from the 16×16 dimension functional array to a 4×4 array of off shell Dirac field functionals. Each component of each off shell Dirac field functional is independent. We define two Dirac field functional 4-vectors

$$f_i(A_i, s) \quad \text{and} \quad f_j(B_j, s) \tag{1.4}$$

where A_i and B_j are sets of internal symmetry indices with $A_i \cup B_j$ form a complete set of indices for i, j = 1, 2, 3, 4. Each functional, $f_i(A_i, s)$ and $f_j(B_j, s)$, has four spin components. Thus F(i, j) below forms a 16×16 array.

$$F(i, j) = f_i(A_i, s) \, f_j(B_j, s)^T \tag{1.5}$$

The indices i, j = 1,2, 3, 4 label 4×4 blocks of functionals.

Thus we can express the 16×16 dimension array as the outer product of two Dirac functional 4-vectors f_i and f_j. We take f_i and f_j to correspond to two more basic Dirac fermion 4-vectors that we call *q-fermions*. The map to a 4×4 array of Dirac functionals is diagrammed in Fig. 1.5.

We can define a 4×4 array of dimensions corresponding to the 4×4 array of functionals as in Fig. 1.5. Then we use the same technique to reduce the basis to one fermion Dirac functional in one dimension as in the lowest part of Fig. 1.5. We call this lower basis the *qq-basis..* This process of reduction makes physical sense because all

QUeST fermions can be transformed to each other. There is a commonality to all fundamental fermions (neglecting interaction effects.)

1.3 Derivation to the qq-basis – A One Fermion Substratum

The q-basis of Dirac 4-vector functionals exists in a 16 dimension space. We view it as a 4×4 space (a 4-space of quaternions).

We can apply the same procedure as above to reduce the space to a one dimension space with one Dirac functional.

We treat each of 4-vector above, f_i and f_j, as generated from four off shell Dirac vectors components. Then a one dimension space follows with one Dirac particle (functional) and corresponding *qq-fermion*.

More directly, we can view the one Dirac functional as part of an 8 dimension Dirac particle, which has 16-spinor components. The outer product of two 8 dimension Dirac particle functionals results in a 16×16 array of 256 dimensions.

An 8 dimension qq-fermion can exist in Megaverse MOST, which has a $7 + 1$ dimension space-time. Then one can consider the universe as beginning as a qq-fermion in the Megaverse and dynamically growing to be the current universe. See section 1.8 below. This view is supported by the derivation of the universe growth rate (Hubble rate) near the Big Bang as a particle vacuum polarization effect. See Blaha (2019e).

All fermions can be generated from any one fermion by applying symmetry transformations. Bearing in mind dimension-fermion duality we can view the dimension dual of the one fermion as the "source" of dimensions. The other 255 dimensions are generated by transformations parallel to the above SU(2)⊗U(1)⊗SU(3) and Complex Lorentz group transformations.

1.4-Dimensional Space-Time

The 4-dimension fermion functional 4-vector of the q-basis maps to 4-dimension fermion vector within a 4-dimension space. We call it *q-space*.

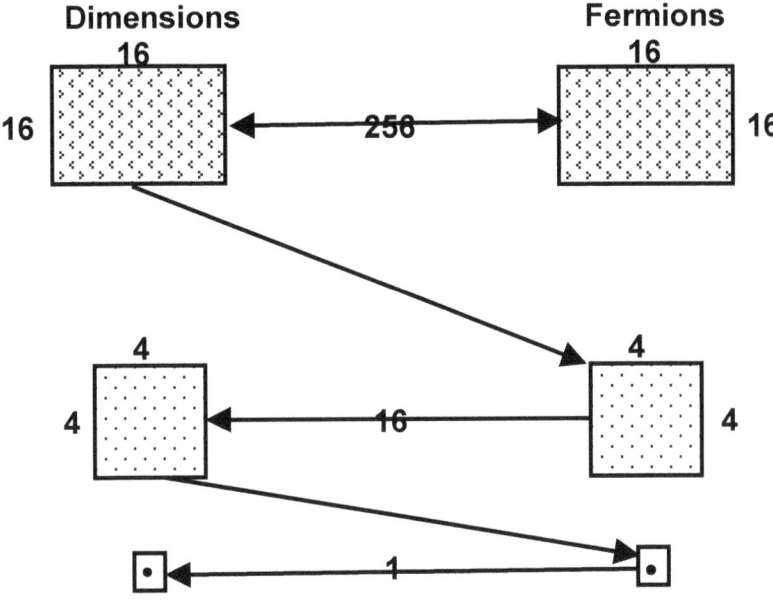

Figure 1.5. Diagram of the descent to the q-basis from QUeST.

1.5 QUeST Fermions are Composites of q-Fermion

The fundamental fermions of QUeST can be viewed as composites of two q-fermions. In particular, a fundamental fermion is composite of a type "1" q-fermion and a type "2" q-fermion.

Physical considerations lead us to specify the composite fermions are tightly bound by a U(4) force between q-fermions. We postulate an ultra-strong, short distance U(4) gauge theory with a dynamics with a momentum space form:

$$G(k) = GM^2/(k^2 - M^2) \qquad \text{as } M \rightarrow \text{infinity} \qquad (1.6)$$

In the limit the potential associated with G(k) becomes a $\delta^3(\mathbf{r})$ potential binding q-fermions together. The constant G has the dimensions of inverse mass squared thus suggesting G is the gravitation constant.

1.6 Megaverse MOST mq-Formulation

MOST is formulated in 32 complex octonion dimensions. The "derivation" of the deeper origin of MOST begins with fermion – dimension duality, in which the number of fermions equals the number of MOST dimensions: 512. This equivalence, which extends down to the individual particle and dimension level, allows us to map between (fundamental) fermions and dimensions. We will call the basis of MOST the mq-basis.[5]

The derivation of MOST from the mq-basis starts with two fermions: a 4-vectors f_1 and an 8-vector-f_2 that we call m*q-fermions*. Using them we define a 4×8 array, which we can express as their outer product:

$$f_1 f_2^T \tag{1.7}$$

where ^{T} again indicates transpose. We define a complex fermion 8-vector from f_2, and use it to define a fundamental representation of U(8).

The 4×8 mq-fermion array generated by eq. 1.7 has the form

$$F_{ij} = f_{1i} f_{2j} \tag{1.8}$$

with $i = 1, 2, \ldots, 4$ and $j = 1, 2, \ldots, 8$. Taking account of the four spinor components of each fermion, which we transform to dimensions as above. We obtain a $16*32 = 512$ component dimensions array D with the dimensions of MOST.

$$D_{ijkl} = f_{1ik} f_{2jl} \tag{1.9}$$

where $k, l = 1, 2, 3, 4$ originate from the spinor components.

The MOST dimensions follow. We can choose to identify the mq-space with the complex coordinate limit of the MOST 8-dimension complex quaternion space-time. There is no reason to treat mq-space as a separate space-time.

---

[5] We use the prefix "mq" to distinguish MOST from QUeST bases.

A disquieting feature of this derivation is the 4-vector nature of f_1. In the next section we define a larger version of MOST that yields 8-vector fermions.

1.7 UTMOST Generalization of MOST

We can define a *64 dimension* complex octonion extension of MOST that we call UTMOST.[6] The "derivation" of the deeper origin of UTMOST again begins with fermion – dimension duality, in which the number of fermions equals the number of MOST dimensions: now 1024. This equivalence again extends down to the individual particle and dimension level, and allows us to map between (fundamental) fermions and dimensions. We will call the basis of UTMOST the *mmq-basis*.[7]

The derivation of UTMOST from the mmq-basis starts with two fermions: an 8-vectors f_1 and another 8-vector-f_2 that we call mm*q-fermions*. Using them we define an 8×8 array, which we can express as their outer product:

$$f_1 f_2{}^T \tag{1.10}$$

where ^{T} again indicates transpose. We define a complex fermion 8-vector using $f_1 + if_2$, and use it to define a fundamental representation of U(8).

The 4×8 mmq-fermion array generated by eq. 1.10 has the form

$$F_{ij} = f_{1i}f_{2j} \tag{1.11}$$

with i, j = 1, 2, ... , 8. Taking account of the four spinor components of each fermion, which we transform to dimensions we obtain a 16*64 = 1024 component dimensions array D—the dimensions of UTMOST.

$$D_{ijkl} = f_{1ik}f_{2jl} \tag{1.12}$$

where k, l = 1, 2, 3,4 originate from transformed spinor indices.

---

[6] See the following chapter.
[7] We use the prefix "mmq" to distinguish UTMOST from MOST.

The UTMOST dimensions follow. The 64 complex octonion dimension UTMOST leads to an 8 dimension complex space that we can take to be equivalent to the 8 dimension octonion space that UTMOST generates when the dimensions are separated into space-time and internal symmetry parts.

We choose to identify the *mmq-space* with the complex coordinate limit of the UTMOST 8-dimension complex octonion space-time. There is no reason to treat mmq-space as a separate space-time. Thus we have a theory that makes fundamental fermions composites of two mmq-fermions. In particular, a fundamental fermion is composite of a type "1" mmq-fermion and a type "2" mmq-fermion.

Physical considerations lead us to suggest (similarly to QUeST) that composite fermions are tightly bound by a U(8) force between mmq-fermions. We postulate an ultra-strong, short distance U(8) gauge theory with a dynamics with a momentum space form:

$$G(k) = GM^2/k^2 - M^2) \qquad \text{as } M \rightarrow \text{infinity} \qquad (1.13)$$

In the limit the potential associated with G(k) becomes a $\delta^3(\mathbf{r})$ potential binding the mmq-fermions together. The constant G has the dimensions of inverse mass squared thus suggesting the G is the gravitation constant. A subsector of mmq-fermions and their forces goes over to the QUeST case in component universes.

1.8 Origin of a 256 Dimension Universe

One might accept the origin of a 4-dimension universe in a Big Bang or otherwise. But the origin of a 256 dimension universe raises serious issues. The ordering of the topology leading to such a universe is a major issue. Also one instinctively feels that large entities should originate in smaller ones. The Big Bang theories espouse this point.

Thus the origin of the QUeST universe with its 256 dimensions raises important issues far beyond conventional Big Bang theories.

A possible mechanism to resolve these issues exists in the one dimension qq-basis where there is only one fermion. The symmetry groups of UST and QUeST can "rotate" the one fermion to all other 255 fermions. Thus we can envision a dynamic origin of the universe from one fermion to the entire gamut of fermions in one instant or longer. A quaternion universe then becomes feasible in principle.

2. Dimension-Fermion SuperSymmetry

The equality of the number of dimensions and of the number of (fundamental) fermions evokes the possibility of SuperSymmetry. As stated in section 1.1 of chapter 1 we can define fields corresponding to each dimension. These fields can be reasonably taken to be scalar quantum fields with the symmetries of QUeST (and UST).[8]

One then has an equal number of spin ½ fermions and spin 0 bosons—fertile ground for the development of a SuperSymmetric formulation. In this chapter we develop a SuperSymmetric extension of QUeST.

2.1 Dimension Particles (Fields) and Functionals

In section 1.1 we showed that dimensions can be naturally interpreted as representing scalar fields containing dimension functionals. The ij-th dimension could be expressed as a functional inner product of a dimension functional f_{ij} and a "wave" in a coordinate space denoted (x):

$$D_{ij} = D_{ij}(x_{ij}) = (f_{ij}, (x_{ij})) \qquad (1.2)$$

The 256 dimension fields $\varphi_i(x)$ can be numbered i = 1, ... , 256.

Consider the 256 dimension functionals in comparison to the 256 fundamental fermion functionals. The 256 dimensions, and dimension fields, of 32-dimension complex quaternion space can be viewed as in fundamental representations of U(256).

However they can also be viewed as composing a product of fundamental representations of

$$G = [SU(2) \otimes U(1) \otimes SU(3) \otimes SU(2) \otimes U(1) \otimes SU(3) \otimes U(4)^4 \otimes U(2)]^4 \qquad (2.1)$$

---

[8] Alternatively, one could define a space with spin ½ dimensions using the equality of dimensions and fermions and hope that this forms the basis of a theory superficially similar to Twistor Theory. Twistor defines spin ½ constructs at every *point* of a space-time. Dimension Twistor Theory or Point Twistor Theory?

plus a 4 dimension complex quaternion space-time, based on a partition into blocks based on the fundamental representation dimension rules:

> U(2) requires 4 real dimensions
> U(1)⊗SU(2) requires 4 real dimensions
> SU(3) requires 6 real dimensions
> U(4) requires 8 real dimensions

where the dimensions have *real-valued* coordinates and are called *real dimensions*. Fig. 2.1 shows the partition of one layer QUeST (with 8 complex quaternion dimensions). Fig. 2.2 shows the partition of four layer QUeST (with 32 complex quaternion dimensions—four copies of the one layer QUeST).

The 256 dimensions of the 32 dimension complex quaternion space is then seen to exactly equal the symmetries of QUeST and UST—a rather remarkable result considering UST was derived by the author several years ago deductively in the manner of Euclidean Geometry.

The 256 dimensions of the 32 dimension complex quaternion space equal the 256 fundamental fermions of QUeST and UST. The number of internal symmetry vector bosons is $(4+8+4+8+64+4) \times 4 = 368$ based on a) 4 vector bosons for SU(2)⊗U(1); b) 8 vector bosons for SU(3); c) 16 vector bosons for U(4); and d) 4 vector bosons for U(2).

2.2 SuperSymmetric Preliminary

The number of dimension functionals equals the number of fermion functionals (256). We can associate both sets of functionals with fields via functional inner products. We choose to use 3 + 1 complex coordinate space in the Fourier representation: (x) = (x_{ij}) in eq. 1.2 to obtain 256 scalar fields $\varphi_i(x)$ for dimensions, and a similar representation for 256 fundamental fermion fields $\psi_{ij}(x)$:

$$\psi_{ij} = \psi_{ij}(x) = (F_{ij}, (x)) \tag{2.2}$$

where F_{ij} represents a fermion functional. We also represent fermion fields by $\psi_i(x)$ for i = 1, … , 256.

The dimension functionals and fermion functionals both provide representations of the internal symmetry groups of eq. 2.1. The group representations of both the scalar dimension particles, and the fermions, are all group fundamental representations.[9]

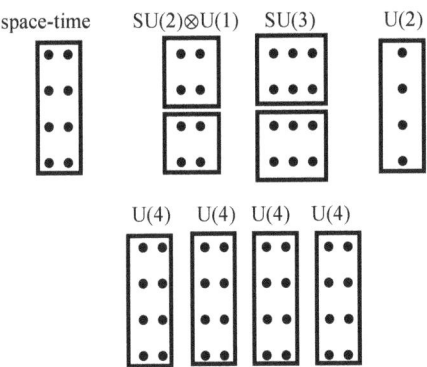

Figure 2.1. Psiphi diagram showing partitioning of 8 dimension complex quaternion space. The blocks of dimensions yield 4 complex dimension space-time, and the internal symmetries of normal matter and Dark matter $SU(2) \otimes U(1) \otimes SU(3) \otimes U(2) \otimes SU(2) \otimes U(1) \otimes SU(3) \otimes U(2)$ as they appear in UST. The lower U(4) groups are for the Generation group and the Layer group for normal matter and for Dark matter in one layer UST. The U(2) group transforms between normal and Dark matter.

---

[9] Thus the problem of standard SuperSymmetric theories that force both vector bosons and fermions into adjoint representations is avoided.

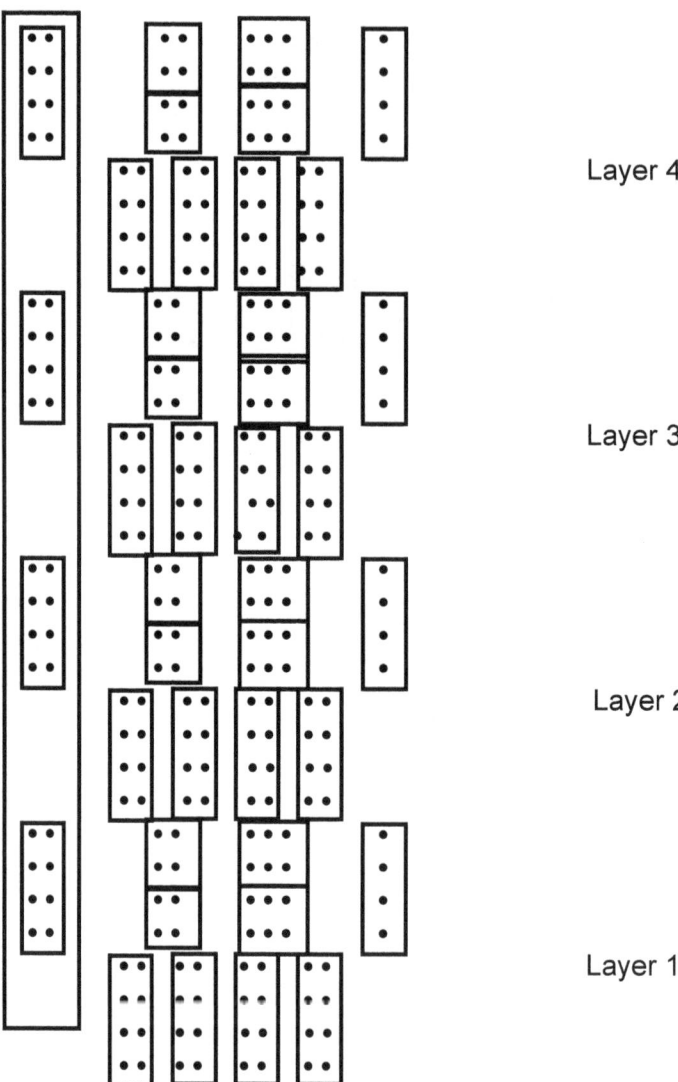

Layer 4

Layer 3

Layer 2

Layer 1

Figure 2.2. Four layer QUeST internal symmetry groups and space-time diagram for 32 dimension complex quaternion space. Note the left composite blocks combine to specify a 4 dimension complex quaternion space-time.

2.3 Θ Transformation Group

An examination of the set of dimensions and the set of fermions in Fig. 1.2 shows a row by row match of dimensions and fermions. However the match is a superficial one. There are 256! possible maps between dimensions and fermions. We could reorder the 256 dimensions in 256! ways independently of the fermion order. Similarly we could independently reorder the fermions. The S_{256} Symmetric group permutations give the necessary reorderings.

It is also possible to transform dimensions using U(256) if we treat the dimensions as a 256-vector in the fundamental representation of U(256). Appling a U(256) transformation to the vector we create a new 256-vector whose components are linear sums of dimensions. The components of the new 256-vector define a new coordinate system as a rotation of that of the previous 256-vector. An analogous transformation is the rotation of an (x, y, z) coordinate system to (x', y', z'). It can be viewed as a rotation of dimensions (i, j, k).

2.3.1 Scalar Dimension Fields

The scalar dimension fields, $\varphi_i(x)$, that are defined for the set of 256 dimensions similarly have a corresponding U(256) transformation group, denoted Θ. The U(256) symmetry group can be taken to be a local Yang-Mills symmetry group.

2.3.2 Fermion Transformations

The 256 fundamental fermions corresponding to the set of dimensions provisionally have a U(256) transformation group, which we denote Θ'. However the breakdown[10] of Θ' to the particle symmetry group G implied by Fig. 2.2 destroys the U(256) symmetry. Thus the fermions have the internal symmetries of G. Θ' transformations are restricted by the charge superselection rule among other things.

The transformations of dimensions and dimension fields are of the Θ group, and are unlike those of Θ' because of symmetry breakdown and conservation laws.

---

[10] At the Big Bang point the symmetry is Θ' = Θ if there is no breakdown.

2.3.3 Relation Between Dimension Fields and Fermion Fields

There is no inherent relation between scalar dimension field symmetry Θ and fermion symmetry Θ'. Transformations of each group are independent. When SuperSymmetric transformations are introduced below, then Θ and Θ' become related although Θ' becomes reduced to G.

2.3.4 Dimension Fields Internal Symmetries

Once the internal symmetries of the fermions and their functionals are specified, then they can be mapped to corresponding dimensions, their dimension fields, and their functionals using the representation in Fig. 1.2.

$$\psi_{ijI}(x) \rightarrow F_{ijI} \rightarrow \varphi_{ijI}(x) \rightarrow f_{ijI} \qquad (2.3)$$

where $\psi_{ijI}(x)$ is the $(ij)^{th}$ fermion with internal symmetries I, F_{ijI} is its functional, $\varphi_{ijI}(x)$ is the $(ij)^{th}$ dimension field with internal symmetries I, and f_{ijI} is its dimension functional.

2.4 SuperSymmetric Transformations

The Θ group transformations of the dimension functionals are independent of the Θ' transformations of the fermion functionals in the free fields case. If we introduce SuperSymmetric transformations between dimensions and fermions, then they are directly related. Unlike the case of conventional SuperSymmetry Dimension-Fermion SuperSymmetry has a purpose.

The G group transformations of both give a symmetry to the dimension fields that enables us to view them as Higgs fields.

2.4.1 Fermion-Dimension SuperSymmetric Transformations

A SuperSymmetric transformation that rotates dimension fields φ and fermion fields ψ is:

$$\delta\varphi_I = \bar{\varepsilon}\,\psi_I \qquad (2.4)$$
$$\delta\psi_I = -i\gamma^\mu\partial_\mu\varphi_I\varepsilon$$

for internal symmetry indices I where ε is a constant anticommuting 4-spinor. The lagrangian

$$\mathscr{L} = i\bar{\psi}_I \gamma^\mu \partial_\mu \psi_I + \partial^\mu \varphi_I \partial_\mu \varphi_I \tag{2.5}$$

is invariant under this transformation.

2.4.2 Fermionic Operator Formalism

One can define two sets[11] of 256^2 fermionic operators $\{A_{\Phi Fnm}(x)\}$ and $\{A_{F\Phi nm}(x)\}$ that transform fermions into scalar dimension fields, and scalar dimension fields into fermions respectively.

$$A_{\Phi Fnm}(x)\psi_m(x) \rightarrow \varphi_n \tag{2.6}$$
$$A_{F\Phi nm}(x)\varphi_m(x) \rightarrow \psi_n$$

where $n = 1, \dots, 256$ and $m = 1, \dots, 256$ label fields irrespective of internal symmetries.[12] Each (n, m) pair labels one of 2×256^2 fermionic operators.

The fermionic operators are closed under anticommutation. One simple 'free' field representation of these operators in terms of the fermionic and bosonic fields is

$$A_{\Phi Fnm}(x) = \varphi_n(x)\psi_m(x) \tag{2.7}$$
$$A_{F\Phi nm}(x) = \psi_n(x)\varphi_m(x)^\dagger$$

where n and m are labels ranging from $1, \dots 256$ for fermions and bosons, $\varphi_n(x)$ is the nth of the 256 dimension fields, $\psi_m(x)$ is the mth of the 256 fermion fields; and $\varphi_m(x)^\dagger$ is the hermitean conjugate of the mth field of the 256 dimension fields and $\psi_n(x)$ is the nth fermion field of the 256 fermion fields. Spatial integrals of the above operators have simpler anticommutation relations.

Elements of the Θ group can be used to map the fermionic operators to different forms. For example if U_Θ is an element of the Θ group[13] we can redefine fermion operators as a different map between fermions and bosons with expressions like

---

[11] While these operators form a Jordan algebra they do not have group representations.

[12] The ability to 'rotate between interactions' using the Θ group enables the fermionic operators to be transformed in a lagrangian covariant manner.

$$\mathbf{A}'_{\Phi Fnm} = U_\Theta \mathbf{A}_{\Phi Fnm} U_\Theta^{-1} \tag{2.8}$$

$$\mathbf{A}'_{F\Phi nm} = U_\Theta \mathbf{A}_{F\Phi nm} U_\Theta^{-1} \tag{2.9}$$

Thus one can SuperSymmetric map between the fermion and dimension fields.

2.4.3 Relation of the Dimension Fields and the Fermionic Fields

The fermionic field operators that we have defined can be related to the U(256) field operators, which we will denote $\mathbf{A}_{\Theta nm}(x)$ where n = 1, ... , 256 and m = 1, ... , 256 are labels on the <u>256</u> fermion or dimension fields. The following relations hold:

1. The product of $\mathbf{A}_{\Phi F}^{\mu}{}_{nm}(x)$ and $\mathbf{A}_{\Phi F}^{\mu}{}_{mp}(x)^\dagger$ can be expressed as a sum of Θ operators.
2. The product of $\mathbf{A}_{F\Phi nm}(x)^\dagger$ and $\mathbf{A}_{F\Phi mp}(x)$ can be expressed as a sum of Θ operators.
3. The product of $\mathbf{A}_{\Phi Fnm}(x)$ and $\mathbf{A}_{F\Phi mp}(x)$ can be expressed as a sum of Θ operators.

Thus we find that the U(256) Θ operators can be obtained from products of the fermionic operators. We also note that the product of a fermionic operator and a Θ operator is a fermionic operator.

These relations establish a close connection between fermionic field operators and the Θ operators.

We conclude that there is a SuperSymmetry relation between the fermion and scalar dimension fields in QUeST. The benefits of SuperSymmetric transformations in this context remain to be determined.

2.4.4 Higgs Dimension Fields

The scalar dimension fields $\psi_I(x)$ may be identifiable with Higgs symmetry breaking fields since they have the same fundamental representations of G as the fundamental fermions. Note that every fermion has a mass. Dimension fields can be defined in four dimension space-time.

---

[13] We let $\Theta = \Theta'$.

3. Probes of Dimensions Outside of Space-Time

The QUeST of our universe has a 32 complex quaternion dimension space with 256 dimensions that contains a 4 complex quaternion space-time, which in turn has the real 3 + 1 space-time of our experience. MOST of the Megaverse has a 32 complex octonion dimension space with 512 dimensions that contains an 8 complex quaternion space-time. In this chapter we outline possible probes that can take us beyond our 3 + 1 space-time.

3.1 Probes into the Internal Symmetry Part of Space

The larger part of QUeST space contains the fundamental representations of the internal symmetry groups of QUeST and UST. Probing this sector appears challenging until we realize that all elementary particle experiments that probe particle interactions are in fact probes of the internal symmetry sector of QUeST space. An examination of the fermion and vector boson sectors of QUeST (and UST) shows there is still much to learn. It appears that the problem lies in the difficulty of creating sufficiently powerful accelerators.

3.2 Probes into the Megaverse

Our universe resides in a larger 7 + 1 dimension Megaverse. We seem to be confined to our universe without extraordinary efforts to emerge. Blaha (2018a) outlines proposals of the author for universe exit via starship slingshots around neutron stars and other very dense, small objects with enormous gravitation fields. The book also suggests a way to exit the universe at "weak spots" where the universe's surface tension is weak. The methods proposed are based on faster-than-light travel, which is possible in UST.

The Essence of Eternity:
Quaternion & Octonion SuperStandard Theories

Stephen Blaha Ph. D.
Blaha Research

Pingree-Hill Publishing
MMXX

Copyright © 2020 by Stephen Blaha. All Rights Reserved.

This document is protected under copyright laws and international copyright conventions. No part of this book may be reproduced, stored in a retrieval system, or transmitted by any means in any form, electronic, mechanical, photocopying, recording, or as a rewritten passage(s), or otherwise, without the express prior written permission of Blaha Research. For additional information send an email to the author at sblaha777@yahoo.com or call 603-289-5435.

ISBN: 978-1-7345834-2-7

This document is provided "as is" without a warranty of any kind, either implied or expressed, including, but not limited to, implied warranties of fitness for a particular purpose, merchantability, or non-infringement. This document may contain typographic errors, technical inaccuracies, and may not describe recent developments. This book is printed on acid free paper.

Rev. 00/00/01   March 26, 2020

INTRODUCTION

In previous books this author has derived the Unified SuperStandard Theory (UST) in our 3 + 1 dimension space-time from Complex General Relativity and Quantum Field Theory suitably extended. This book expands on Blaha (2020c) to give a more detailed derivation of UST from a 32 dimension complex quaternion space. Remarkably the internal symmetries of UST follow directly from the complex quaternion space discussion.

Similarly it also defines a complex octonion space of 32 dimensions that leads to an acceptable Megaverse with our universe embedded within it.

The general conclusions of our development are:

Our universe is a 32 dimension Complex Quaternion Space. It is equivalent to QUeST, which has a 3 + 1 Complex Quaternion Dimensions Space-Time and UST Internal Symmetries.

Upon restriction of QUeST to real-valued coordinates, QUeST Becomes the Unified SuperStandard Theory (UST).

The Megaverse is a 32 dimension Complex Octonion Space. It is equivalent to MOST, which has a 7 + 1 dimensions Complex Quaternion Space-Time and Internal Symmetries.

The 7 + 1 dimensions Complex Quaternion Space-Time of the MOST Megaverse contains 3 + 1 dimensions Complex Quaternion QUeST Universes' Space-Times.

QUeST → real-valued coordinates Unified SuperStandard Theory

We show that the Unified SuperStandard Theory is an exact consequence of Quaternion and Octonion space. *The generality of these results, and the author's previous derivation of the exact value of the Fine Structure Constant α, and approximate derivations of the Weak and Strong coupling constants (shown later), shows that the*

Internal Symmetries of the Standard Model, and our Unified SuperStandard Theory, are now understandable.

Thus particle physics can be explained by Quaternion and Octonion space characteristics. We show this in some detail in the following chapters.

1. Extension of the Unified SuperStandard Theory (UST) to Higher Dimensions

1.1 Suggestions of a Deeper Basis for UST

The UST provides a complete theory of elementary particles and Gravitation in 3 + 1 dimensions (real-valued coordinates). It is presented in detail in Blaha (2020c). In the study of this theory extending back 7+ years to Blaha (2012a) and earlier books, the author found a close analogy between the subgroups of the Complex Lorentz Group and the factors of the Standard Model internal symmetry: $SU(2) \otimes U(1) \otimes SU(3)$.

This seeming coincidence raised the question that the analogy was based on a deeper relation embodied in a larger space. This larger space would have a subspace devoted to space-time and a subspace for internal symmetry groups. The subspaces would be orthogonal.

Pursuing this concept lead the author, in the fall, 2019, to develop a formulation of larger spaces based on hypercomplex coordinates: quaternions and octonions. This development reached fruition in Blaha (2020a), (2020b) and (2020c)..

1.2 Hypercomplex Numbers

The key concept in the search for a deeper basis for UST was the following progression of coordinate system formulations:

A. The Standard Model was based on real-valued coordinates.

B. UST is based on real-valued coordinates made complex:

B1. Real-valued coordinates were replaced by slightly complex coordinates with an imaginary part consisting of a massless second quantized vector field that gave the usual perturbation theory results at low energy and \eliminated divergences at very high energy

B2. Quarks and Strong Interaction gauge fields were truly complex (with certain restrictions) due to Complex Lorentz Group boosts.

C. Having progressed from real-valued coordinates to complex-valued coordinates in parts A and B, we considered hypercomplex-valued coordinates and developed 32 complex quaternion coordinate systems and 32 dimension complex octonion coordinate systems.[14] We suggested complex quaternion coordinates were appropriate for our universe, and complex octonion coordinate systems were appropriate for a Megaverse (Multiverse) should our universe reside in it. The author found that the complexity of quaternion space and octonion space was important for the analogy with complex space-time of UST. This is discussed in Blaha (2020c) and later in this book.

D. The choice of 32 dimension coordinate systems was motivated by the perceived need to have a complex quaternion space-time for universes and a complex octonion space-time for the Megaverse. A benefit of this approach was that it forced layers of particles to exist in the resulting theories—consistent with the layers found in the real-valued coordinates UST.

The result of this chain of progression was a complex quaternion space-time theory (called QUeST) that directly becomes UST upon restriction to real-valued coordinates. Further, upon restriction of the complex octonion space-time theory (called MOST) to a complex quaternion space-time we obtain QUeST. Thus there are three closely related theories UST, QUeST, and MOST. We discuss details related to this section later.

1.3 Mathematical Picture Language

As we will see later we need only consider the allocation of dimensions to space-time, and to the various internal symmetry groups in the 32 dimension spaces. The detailed mathematics of quaternions and octonions is not directly relevant for the internal symmetry subspaces. The quaternions and octonion coordinates of the space-times can utilize the algebras of these hypercomplex numbers in calculations.

Since the allocation[15] of dimensions to fundamental representations of groups is of paramount importance to establish the basis of UST we simply used the symbol • to represent a dimension. The many dimensions of the 32 dimension spaces, which take account of the dimensions within a quaternion quartet, and an octonion octet, are thus represented by patterns of •'s. The separation of the dimensions (•'s) into fundamental representations is then easily accomplished.

---

[14] The use of quaternion and octonion number systems is predicated on the nature of S matrix elements, which necessarily use the Dyson-Wick expansion of time ordered products in perturbation theory. S matrix elements require the rules of real numbers or complex numbers or quaternion numbers or octonion numbers for calculations.

[15] The allocation of dimensions is a euphemism for the use of transformations to block diagonalize the 32 dimension spaces.

Thus we have a simple Mathematical Picture Language for the separation that would require complex transformations otherwise as indicated in the footnote below.

1.4 Number Describes Everything

Pythagoras developed a language of •'s later for computation. He called the •'s *psiphi* meaning pebbles in Classical Greek. In this regard he developed a set of diagrams for numbers. The perfect number 10 was represented by

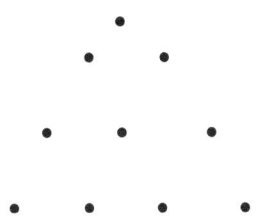

Figure 1.1. Pythagorean tetrackys, a psiphi representation of 10.

He considered ten to be the perfect number because it embodies 1, 2, 3, and 4. Remarkably perhaps, the groups U(1), U(2), SU(3), and U(4) are the internal symmetry groups of UST and the Standard Model.

Pythagoras' great achievements and his innate sense of the nature of Reality led to the proposition: Our theories directly reflect its truth.

$$\text{\textbf{The essence of all things is number.}} \tag{1.1}$$

From total symmetry of the internal symmetry subspace, and the use of the number of dimensions of fundamental representations of internal symmetry groups, we are led directly to the internal symmetry groups, the fermion spectrum, the vector boson spectrum and Gravitation.

We will see "blocks" of fundamental representations of the groups known from the Standard Model (and UST) appearing in the quaternion and octonion theories:

$$SU(2)\otimes U(1)\otimes SU(3)\otimes SU(2)\otimes U(1)\otimes SU(3) \tag{1.2}$$

where the second set of $SU(2)\otimes U(1)\otimes SU(3)$ factors are symmetries of Dark matter in UST. The total dimension of these factors is 10, the Pythagorean perfect number, which can be viewed as an interesting coincidence. It suggests the Pythagorean summation of Reality should be:

Number is the Essence of Eternity   (1.3)

since one can hope the theories, UST, QUeST, and MOST, if correct, are eternal being all based on Number.

1.5 Axioms for the Unified SuperStandard Theories

We list a set of Axioms for the combined Unified SuperStandard Theories for our 3 + 1 dimension universe, for the 3 + 1 complex quaternion QUeST universe, and for the 7 + 1 octonion MOST Megaverse.

AXIOMS

1. A biquaternion space is the basic space-time of our universe. Bioctonion space is the basic space-time of the Megaverse.

2. Physical processes can execute in parallel.

3. Matter and energy are particulate.

4. Space--times are locally Lorentzian.

5. All calculations are finite.

6. Particle theory can be defined in any curved space-time.

7. Each particle has a wave function determined by a functional inner product

8. defining the particle state. The functionals form a set without a distance measure.

1.6 General Implications of the Axioms

In this section we describe some of the implications of each of the axioms.

1. Biquaternion space is the basic space-time of our universe. Bioctonion space is the basic space-time of the Megaverse.

The factorization into a space-time and an internal symmetry space must be a form of spontaneous symmetry breaking of yet unknown origin. It appears to be related to a breakdown of the vacuum.

2. Physical processes can execute in parallel.

Physical processes are known to be able to execute in parallel at any distance of separation. As Fant has shown parallel execution requires a minimal number of dimensions: 4. Consequently the dimension of space-time must be 4 or greater. The biquaternion space-time of QUeST is 4-dimensional allowing parallel process execution.

The bioctonion space-time of MOST is 8-dimensional and also allows parallel process execution. The choice of eight dimensions is natural since it allows 4-dimensional universes within it. It also has a form that allows a clean formulation. Lastly, as will be seen later, it conforms to the pattern of interplay between Lorentz symmetry and internal symmetry found in the Unified SuperStandard Theory. This axiom leads to a view of the origin of the dimensions.

3. Matter and energy are particulate.

The most direct method of specifying a theory of matter and energy is through the Use of Quantum Field Theory. Thus Quantum Field Theory is implied.

4. Complex Space-times are locally Lorentzian.

A locally complex Lorentzian space-time leads to Complex General Relativity. In flat space-time Complex General Relativity becomes Complex Lorentz group. (In point of fact the Complex Poincaré group follows.)

5. All calculations are finite.

Given the need for Quantum Field Theory it becomes necessary to find a formulation that yields finite values for calculations in perturbation theory. The only approach that eliminates high energy divergences, and yet preserves the results found in perturbation theory calculations that agree with (primarily QED) experiments, is Two-Tier Quantum Field Theory. This is discussed in detail in earlier books starting in 2002. Thus only our Two-Tier formalism satisfies this axiom.

6. Particle theory can be defined in any curved space-time.

In the 1970s we developed a formalism that allows the definition of particle states in any space-time in such a way that its physical content is preserved when transformed to any coordinate system.[16] This PseudoQuantum Quantum Field Theory satisfies this axiom.

7. Each particle has a wave function determined by a functional inner product defining the particle state. The functionals form a set without a distance measure.

This axiom is satisfied by our formulation of quantum functionals in Blaha (2019f) and earlier books. Our formulation eliminates the superficial violation of the Theory of Relativity by "spooky" quantum entangled processes with parts separated by a physically "large" distance.

The seven axioms imply the Unified SuperStandard Theory and its deeper biquaternion and bioctonion hypercomplex formulations.

---

[16] S. Blaha, Il Nuovo Cimento **49A**, 35 (1979).

2. Hypercomplex Coordinate Systems

2.1 Hypercomplex Number Based Higher Dimensions

The Unified SuperStandard Theory was based on real-valued and complex-valued coordinates. This choice enabled us to understand the reason behind the four types of fermions found in nature: neutral fermions, charged fermions, up-type quarks, and down-type quarks. A study of Complex Lorentz group subgroups showed that they were similar to the factors of the Standard Model symmetry group. This similarity motivated this author to consider a space that integrated space-time and internal symmetry dimensions.

In choosing a higher dimension space for a larger theory of elementary particles the use of coordinate systems based on hypercomplex number systems seemed reasonable. The hope was that just as complex coordinates led to a deeper understanding of the internal symmetries of the Standard Model, the use of hypercomplex coordinates might lead to a further understanding of the origin of both space-time and internal symmetries.

The use of quaternion and octonion coordinate systems was motivated by the nature of S matrix elements, which necessarily use the Dyson-Wick expansion of time ordered products in perturbation theory.[17] S matrix element calculations require the arithmetic rules of real numbers or complex numbers or quaternion numbers or octonion numbers for calculations.

The pattern of rising hypercomplexity is:

Real → Complex → Quaternion → Biquaternion → Octonion → Bioctonion

The Unified SuperStandard Theory took particle theory from real-valued coordinates to complex-valued coordinates. Complex quaternion (biquaternion) and complex octonion (bioctonion) extensions took us to QUeST and MOST. Both use larger spaces to unite space-time symmetry and internal symmetry.

---

[17] Time-ordered products are defined in quaternion perturbation theory in terms of a "time" defined as a quaternion inner product of a chosen direction in quaternion time, and the time quaternion. Quaternion time is thus single-valued. Potential infinities in higher dimension perturbation theory are eliminated using the author's Two Tier formulation of Quantum Field Theory. See Blaha (2005a).

Since the requirement of parallel physical processes[18] made the minimal space-time dimension 4 and since the Megaverse must include universes as subspaces, we were led to a 4-dimensional complex quaternion formulation for our universe and an 8-dimensional complex octonion formulation for the Megaverse. They were considered in Blaha (2020a), (2020b) and Blaha (2020c). Generalizations to higher dimensional quaternion and octonion spaces were considered to accommodate the four layers suggested by the Unified SuperStandard Theory.

In Blaha (2020c) we considered these higher dimensional theories and found that the complex quaternion higher dimensional theory QUeST) leads directly to the 3 + 1 dimensional Unified SuperStandard Theory upon restriction of the theory to real-valued coordinates. Similarly the Megaverse MOST leads to a reasonable Megaverse theory upon restriction to a generalization of the Unified SuperStandard Theory in 7 + 1 real space-time dimensions.

2.2 From Complex Coordinates to Hypercomplex Coordinates

In this section we show a natural generalization of coordinate systems to hypercomplex number systems. There are a number of ways to define hypercomplex coordinate systems. We extrapolate from complex coordinates:

$$t = t_1 + it_2$$
$$x = x_1 + ix_2$$
$$y = y_1 + iy_2$$
$$z = z_1 + iz_2$$
$$\dots$$

Note that the real part of each coordinate is the same type as the imaginary part—also a real number.

2.3 Quaternion Coordinate Systems

Based on that simple fact it seems natural to define a 3 + 1 dimensional "real-valued" space with quaternion coordinates as:

Time quaternion
$$t = (a + ib + jc + kd) \tag{2.1}$$
Spatial quaternions
$$x = (a_x + ib_x + jc_x + kd_x)$$
$$y = (a_y + ib_y + jc_y + kd_y)$$

---

[18] See Blaha (2020c).

$$z = (a_z + ib_z + jc_z + kd_z)$$

where a, b, c, d are real-valued numbers. The symbols i, j, and k are fundamental quaternion units. Quaternions are described in Appendix 2-A. A quaternion embodied four coordinates. Thus it is a 4-dimensional entity and we attribute 4 dimensions to each quaternion and 8 dimensions to each complex quaternion.

In the UST development (Blaha (2020c)) we saw that complex coordinates and the Complex Lorentz Group were needed to find the four types (species) of fundamental fermions[19] and the internal symmetry groups. Consequently we will use complex quaternions (biquaternions) to develop QUeST, initially in a 7 + 1 complex quaternion space.

Time Biquaternion
$$t = (a + ib + jc + kd) + I(a' + ib' + jc' + kd') \qquad (2.2)$$

Spatial Biquaternions
$$x = (a_x + ib_x + jc_x + kd_x) + I(a'_x + ib_x' + jc_x' + kd_x')$$
$$y = (a_y + ib_y + jc_y + kd_y) + I(a'_y + ib_y' + jc_y' + kd_y')$$
$$z = (a_z + ib_z + jc_z + kd_z) + I(a'_z + ib_z' + jc_z' + kd_z')$$
$$x1 = (a_{x1} + ib_{x1} + jc_{x1} + kd_{x1}) + I(a'_{x1} + ib_{x1}' + jc_{x1}' + kd_{x1}')$$
$$y1 = (a_{y1} + ib_{y1} + jc_{y1} + kd_{y1}) + I(a'_{y1} + ib_{y1}' + jc_{y1}' + kd_{y1}')$$
$$z1 = (a_{z1} + ib_{z1} + jc_{z1} + kd_{z1}) + I(a'_{z1} + ib_{z1}' + jc_{z1}' + kd_{z1}')$$
$$w1 = (a_{w1} + ib_{w1} + jc_{w1} + kd_{w1}) + I(a'_{w1} + ib_{w1}' + jc_{w1}' + kd_{w1}')$$

where I is an additional fundamental quaternion unit. Note that the real and imaginary part of each coordinate has the same fundamental quaternion units to permit complex rotations between them.

As we will see we need four iterations of the above set of complex quaternions for the space of QUeST so that it will yield UST upon restriction to real-valued coordinates. Thus QUeST requires a 32 dimension complex quaternion space:

$$t = (a + ib + jc + kd) + I(a' + ib' + jc' + kd') \qquad (2.3)$$
$$x = (a_x + ib_x + jc_x + kd_x) + I(a'_x + ib_x' + jc_x' + kd_x')$$
$$y = (a_y + ib_y + jc_y + kd_y) + I(a'_y + ib_y' + jc_y' + kd_y')$$
$$z = (a_z + ib_z + jc_z + kd_z) + I(a'_z + ib_z' + jc_z' + kd_z')$$
$$x1 = (a_{x1} + ib_{x1} + jc_{x1} + kd_{x1}) + I(a'_{x1} + ib_{x1}' + jc_{x1}' + kd_{x1}')$$
$$y1 = (a_{y1} + ib_{y1} + jc_{y1} + kd_{y1}) + I(a'_{y1} + ib_{y1}' + jc_{y1}' + kd_{y1}')$$

---

[19] Charged leptons, neutral leptons, up-type quarks, and down-type quarks.

$$z1 = (a_{z1} + ib_{z1} + jc_{z1} + kd_{z1}) + I(\,a'_{z1} + ib_{z1}' + jc_{z1}' + kd_{z1}')$$
$$w1 = (a_{w1} + ib_{w1} + jc_{w1} + kd_{w1}) + I(\,a'_{w1} + ib_{w1}' + jc_{w1}' + kd_{w1}')$$
$$\cdots$$
$$w4 = (a_{w4} + ib_{w4} + jc_{w4} + kd_{w4}) + I(\,a'_{w4} + ib_{w4}' + jc_{w4}' + kd_{w4}')$$

Our primary interest in quaternion space is its dimensions and how they can be transformed into a space-time part and an internal symmetry part. Since this purpose is best approached by simply displaying a Mathematical Picture Language representation using Pythagoras' *psiphi* (pebbles) diagrams where each dimension is represented by an • in Fig. 2.1.

The real dimensions[20] of the 7 + 1 space of eq. 2.2 number 64. The real dimensions of the 7 + 1 space of eq. 2.3 number 256. In chapter 3 we will select subsets of dimensions identifying each subset with the fundamental representation of a group. *Thus the spaces of eqs. 2.2 and 2.3 specify space-time and internal symmetry groups numerically. Then the internal symmetries specify the fermion and vector boson spectrums. Thus the consideration of quaternion spaces, which leads to UST and QUeST, devolves into Number as conjectured by the Pythagorean School.*

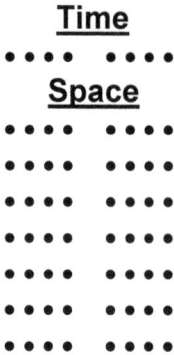

Figure 2.1. Psiphi diagram of the dimensions of 7 + 1 complex quaternion space. Each row represents a complex quaternion. Note the coordinates of eq. 2.2 map directly to dimensions represented by •'s. Each complex quaternion embodies 8 dimensions.

---

[20] Real-valued coordinates are said to have real dimensions. Complex coordinates have complex dimensions.

2.4 Octonion Coordinate Systems

The use of octonion coordinate systems is consistent with the calculation of S matrix elements using the Dyson-Wick expansion of time ordered products in perturbation theory.[21] S matrix element calculations require the arithmetic rules of real numbers or complex numbers or quaternion numbers or octonion numbers for calculations. Octonions can be used to define a coordinate system. An octonion has 8 coordinates and thus it has dimension 8. See Appendix 2-B for a brief description of octonions.

We can define a 3 + 1 dimensional real octonion space as:

Time Octonion
$$t = a + ib + jc + kd + h'a' + i'b' + j'c' + k'd' \qquad (2.4)$$
Spatial Octonions
$$x = a_x + ib_x + jc_x + kd_x + h'a'_x + i'b_x' + j'c_x' + k'd_x'$$
$$y = a_y + ib_y + jc_y + kd_y + h'a'_y + i'b_y' + j'c_y' + k'd_y'$$
$$z = a_z + ib_z + jc_z + kd_z + h'a'_z + i'b_z' + j'c_z' + k'd_z'$$

where a, b, c, d, a', b', c', d' are *real-valued* numbers. The symbols i, j, k, h', i', j' and k' are fundamental octonion units.

In the MOST development (Blaha (2020c)) we saw that complex octonions were needed initially in a 7+ 1 dimensions complex octonion space, which was subsequently expanded four-fold to a 32 dimension complex octonion space. The 7 + 1 dimensions complex octonion space coordinates are:

Time Octonion
$$t = a + ib + jc + kd + h'a' + i'b' + j'c' + k'd' \qquad (2.5)$$
Spatial Octonions
$$x = a_x + ib_x + jc_x + kd_x + h'a'_x + i'b_x' + j'c_x' + k'd_x'$$
$$y = a_y + ib_y + jc_y + kd_y + h'a'_y + i'b_y' + j'c_y' + k'd_y'$$
$$z = a_z + ib_z + jc_z + kd_z + h'a'_z + i'b_z' + j'c_z' + k'd_z'$$
$$x1 = a_{x1} + ib_{x1} + jc_{x1} + kd_{x1} + h'a'_{x1} + i'b_{x1}' + j'c_{x1}' + k'd_{x1}'$$
$$y1 = a_{y1} + ib_{y1} + jc_{y1} + kd_{y1} + h'a'_{y1} + i'b_{y1}' + j'c_{y1}' + k'd_{y1}'$$
$$z1 = a_{z1} + ib_{z1} + jc_{z1} + kd_{z1} + h'a'_{z1} + i'b_{z1}' + j'c_{z1}' + k'd_{z1}'$$

---

[21] Time-ordered products are defined in octonion perturbation theory in terms of a "time" defined as a octonion inner product of a chosen direction in octonion time, and the time octonion. Octonion time is thus single-valued. Potential infinities in higher dimension perturbation theory are eliminated using the author's Two Tier formulation of Quantum Field Theory. See Blaha (2005a).

$$w1 = a_{w1} + ib_{w1} + jc_{w1} + kd_{w1} + h'a'_{w1} + i'b_{w1} + j'c_{w1}' + k'd_{w1}'$$

where all coefficients: a, b, c, d, a′, b′, c′, d′, and a_i, b_i, c_i, d_i, a'_i, b'_i, c'_i, d'_i for i = x, y, z, w, x1, y1, z1, w1 are *complex-valued* numbers. Each complex octonion embodies 16 dimensions. See Fig. 2.2 for the psiphi representation of dimensions in eq. 2.5.

As we saw in Blaha (2020c) we need four iterations of the above set of complex octonions for the space of MOST for reasons discussed later. Thus MOST requires a 32 dimension complex octonion space. The number of real dimensions in this space is 512. The coordinates of 32 dimension complex octonion space are:

Time Octonion
$$t = a + ib + jc + kd + h'a' + i'b' + j'c' + k'd' \qquad (2.6)$$
Spatial Octonions

$$x = a_x + ib_x + jc_x + kd_x + h'a'_x + i'b_x' + j'c_x' + k'd_x'$$
$$y = a_y + ib_y + jc_y + kd_y + h'a'_y + i'b_y' + j'c_y' + k'd_y'$$
$$z = a_z + ib_z + jc_z + kd_z + h'a'_z + i'b_z' + j'c_z' + k'd_z'$$
$$x1 = a_{x1} + ib_{x1} + jc_{x1} + kd_{x1} + h'a'_{x1} + i'b_{x1}' + j'c_{x1}' + k'd_{x1}'$$
$$y1 = a_{y1} + ib_{y1} + jc_{y1} + kd_{y1} + h'a'_{y1} + i'b_{y1}' + j'c_{y1}' + k'd_{y1}'$$
$$z1 = a_{z1} + ib_{z1} + jc_{z1} + kd_{z1} + h'a'_{z1} + i'b_{z1}' + j'c_{z1}' + k'd_{z1}'$$
$$w1 = a_{w1} + ib_{w1} + jc_{w1} + kd_{w1} + h'a'_{w1} + i'b_{w1}' + j'c_{w1}' + k'd_{w1}'$$

$$\cdots$$

$$w4 = a_{w4} + ib_{w14} + jc_{w4} + kd_{w4} + h'a'_{w4} + i'b_{w4}' + j'c_{w4}' + k'd_{w4}'$$

where all coefficients: a, b, c, d, a′, b′, c′, d′, and a_i, b_i, c_i, d_i, a'_i, b'_i, c'_i, d'_i for i = x, y, z, w, x1, y1, z1, w, … , w4 are *complex-valued* numbers.

For the Megaverse we will (in chapter 7) select subsets of dimensions identifying each subset with the fundamental representation of a group. *Thus the spaces of eqs. 2.5 and 2.6 specify space-time and internal symmetry groups numerically. Then the internal symmetries specify the fermion and vector boson spectrums. Thus the consideration of octonion spaces leads to MOST, which devolves into Number as conjectured by the Pythagorean School. All is Number!*

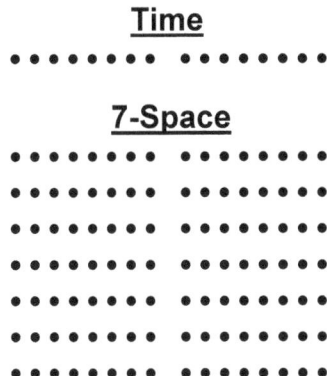

Figure 2.2. Psiphi diagram of the dimensions of 7 + 1 complex octonion space. Each row represents a complex octonion. Note the coordinates of eq. 2.2 map directly to pairs of dimensions represented by •'s. Each complex octonion embodies 16 real dimensions.

Appendix 2-A. Quaternion Features

Quaternions and octonions are hypercomplex numbers with special properties that make them similar to complex numbers.[22] Quaternions and octonions are both normed division algebras over the reals (*hypercomplex* number systems) with salutary properties for quantitative studies in quantum field theory and perturbation theory. Some of their new features are listed on the cover page.

Quaternions have significant properties that distinguish them:

1 .They are associative.

2. They are one of the two finite dimensional division rings having the real numbers as a proper subring. (The other is octonions—considered in chapter 8.)

3. They are non-commutative. (This is not a roadblock for quantum field theory which is also non-commutative in general.)

These features support the development of physics theories.[23]

2-A.1 Some Basic Quaternion Features

A quaternion is a 4-tuple of real numbers. A complex quaternion is a 4-tuple of complex numbers:

$$x = a + bi + jc + kd \ = a + \mathbf{v} \qquad (2\text{-A.1})$$

where a, b, c, d are real or complex numbers, and $\mathbf{v}$ is a 3-vector. The symbols i, j, and k are fundamental quaternion units. A quaternion norm is defined by

$$\|\mathbf{x}\| = sqrt(aa^* + bb^* + cc^* + dd^*) \qquad (2\text{-A.2})$$

and the norm of $\mathbf{v}$ is

$$\|\mathbf{v}\| = \ sqrt(bb^* + cc^* + dd^*) \qquad (2\text{-A.3})$$

An important identity is

---

[22] Much of this appendix appears in Blaha (2020a) and (2020b).

[23] There is an extensive literature on quaternions starting with the original work of Hamilton. Some recent, relevant papers are: S. L. Adler, "Generalized Quantum Dynamics", IASSNS –HEP-93/32 (1993); S. De Leo, arXiv:hep-th/9506179 (1995); Rolf Dahm, arXiv:hep-th/9601207 (1996); S. De Leo, arXiv:hep-th/9508011 (1995); S. L. Adler, arXiv:hep-th/9607008 (1996) and references therein.

$$e^x = e^a \left(\cos (\|\mathbf{v}\|) + \mathbf{v}/\|\mathbf{v}\| \, \sin(\|\mathbf{v}\|) \right) s \qquad (2\text{-}A.4)$$

.It is used to define boosts in quaternion space.

Appendix 2-B. Some Octonion Features

Octonions have significant properties that enable them to be used in a quantum field theory development:

1. An octonion is an 8-tuple of real numbers. A complex octonion is an 8-tuple of complex numbers.
2. They are nonassociative.
3. They are one of the two finite dimensional division rings having the real numbers as a proper subring. (The other is quaternions—considered in chapter 5.)
4. They are non-commutative. (This is not a roadblock for quantum field theory which is also non-commutative in general.)

These features support the development of physics theories.

We can represent a complex octonion (bioctonion) b as

$$b = b_{real} + Ib_{imaginary}$$

where b_{real} and $b_{imaginary}$ are real-valued octonions. We can also represent a complex octonion as an octonion with complex coordinates as in eq. 2.5.

3. Quaternion Unified SuperStandard Theory (QUeST)

In Blaha (2020a), (2020b) and (2020c) we developed QUeST. This chapter expands on those books.

3.1 Motivation and Procedure

Our initial goal was to create a larger dimension space within which we can derive our space-time and the Unified SuperStandard Theory in such a way as to understand the analogy between Lorentz subgroups and Standard Model internal symmetry groups that we found in Blaha (2018f) and earlier books. We expected the development of a deeper form of the Unified SuperStandard Theory (UST) to lead to refinements.

We found some significant new features—most importantly U(2) groups that transformed between normal and Dark matter. These groups must have extremely large mass vector bosons and/or extremely weak coupling constants to be in accord with experimental observations.

There are two possible procedures to follow in developing the deeper basis:

1. One can develop the Quantum Mechanics and Quantum Field Theory in a quaternion space and then extract the dynamics, fermion spectrum, gauge fields, and so on of our familiar space-time.

2. One can define a quaternion space, and then using its coordinates, directly extract the space-time, internal symmetries, fermion spectrum, gauge field spectrum and dynamics. Quaternion algebra may not be used initially. A quaternion space serves to give shape to the range of possible coordinate systems.

We chose the latter approach as it led directly to the Unified SuperStandard Theory in 3 + 1 real-valued dimensions.[24]

In developing the deeper space, upon which we built, we took guidance from the derivation of the Unified SuperStandard Theory. That theory assumes a complex 4-

---

[24] This was an unexpected, but welcome, surprise for the author.

dimensional space-time upon which Complex General Relativity is constructed. It then proceeds to complex flat space-time and Complex Relativity. Next it restricts the complex coordinates of the theory to real-valued coordinates (excepting quarks and color gluons that remain complex as described in Blaha (2020c)). The four types of fermions followed from Complex Lorentz boosts and the internal symmetries reflected the subgroups of the Complex Lorentz Group.

3.2 Definition of A Complex Quaternion (Biquaternion) Space

Following the stated procedure we defined a complex quaternion (biquaternion) space with one "time" biquaternion and seven "spatial" biquaternions believing the 3+1 space-time of our experience is a consequence of this deeper level. Eq. 2. 2 has the 7 + 1 complex quaternion coordinate system.

We do not use the algebra of quaternions but simply treat the quaternion coordinates as coordinates in a space. Our only need was to determine internal symmetries based on the dimensions of fundamental *group representations: 4 real dimensions for SU(2)⊗U(1); 6 real dimensions for SU(3); and 8 real dimensions for U(4).*

Fig 2.1 symbolically depicts the dimensions of the 7 + 1 complex quaternion space. This space has 64 real dimensions (32 complex dimensions.).

3.2.1 Dimension Functionals

The dimension "array" in Fig. 2.1 can be represented in terms of *Dimension Functionals*. We can define a dimension row functional D_i that takes a column dimension as its argument and generate the dimension array $[d_{ij}]$ of Fig. 2.1 with

$$d_{ij} = D_i(d_j) \qquad (3.1)$$

where $i = 1, 2, ..., 8$ and $j = 1, 2, ---, 8$. A dimension column functional may be defined that enables a dimension array to be generated by a functional E_j from a single initial dimension:

$$d_j = E_j(d) \qquad (3.2)$$

Combining eqs. 3.1 and 3.2 we may generate a dimension array with

$$d_{ij} = D_i(E_j(d)) \qquad (3.3)$$

from a single dimension d. We can envision a universe created from a single dimension using this formalism and a suitable dynamics.

Dimension functionals can also be used in the case of octonion coordinates as seen later.

This formulation illustrates the use of dimensions as functionals. We will discuss functionals later where we map dimensions to particle functionals.

3.3 Biquaternion Lorentz Group

Our definition of time and space biquaternion coordinates purposefully resembles those of our real space-time. One might ask why should there be a Lorentz-like group for biquaternion space. The only apparent reason is the need for a special speed c in our space-time that enables one to boost from a rest frame of a mass m particle with energy m to a moving frame of greater energy.[25] Without c the group of the above coordinates would presumably be the biquaternion U(8) group. In this group, transformations preserve the norm of a state so that a "boost-like" transformation does not exist.

The biquaternion Lorentz transformations do have a unique speed c (the speed of light) and specify a unique rest frame for any particle—both sublight particles and tachyon particles.[26] Thus we select the Biquaternion Lorentz group SU(7, 1) for 7 + 1 dimension biquaternion space.

Flat space biquaternion Special Relativity generalizes directly to a biquaternion General Relativity which may be constructed directly (mindful of quaternion non-commutativity).

The flat space biquaternion Lorentz group transformations have constant biquaternion matrix elements that are analogous to those of the Lorenz group.. (See Blaha (2020c) for Lorentz group boosts.)

Motions in complex quaternion space are of four types: motions where the norm of the velocity is real-valued and less than c; motions where the norm of the velocity is real-valued and greater than c; motions where the norm of the velocity is complex-valued and whose absolute value is less than c; and motions where the norm of the velocity is complex-valued and whose absolute value is greater than c,

Boosts generating these cases lead to the separation of fermions into four species for the Unified SuperStandard Theory and its complex quaternion basis QUeST (as well the complex octonion Megaverse MOST.)

---

[25] A similar consideration applies to tachyons.
[26] Tachyons can be transformed by a Complex Lorentz transformation to and from a rest frame. See Blaha (2018f).

3.4 Extracting the Symmetries and Particle Spectra

As stated earlier in section 3.2 we will directly describe the symmetry structure implied by the form of the biquaternion coordinate system while mindful of section 3.3.

The Unified SuperStandard Theory developed the group structure from which the particle species were derived from a subset of Lorentz boost transformations. It used boosts with complex exponentiation similar to quaternion exponentiation in eq. 2-A.4. Complex boosts mapped a system at rest to a system in motion with a real energy and complex 3-momenta in general. Biquaternion boosts play a similar role.

The 4-dimensional subspace for the Unified SuperStandard Theory complex coordinate boosts' was

Time

•

Space

• •

• •

• •

Figure 3.1. Four-Dimensional subspace for Unified SuperStandard derivation of particle spectra with coordinates represented by psiphi • 's. Note the time coordinate is real-valued.

Following the same line of reasoning we now specify a biquaternion subspace analogously restricted to that of Fig. 3.1 to define the relevant set of coordinates for determining particle symmetries and spectra. Note time is a "real" quaternion in this subspace. The spatial coordinates consist of complex quaternions.

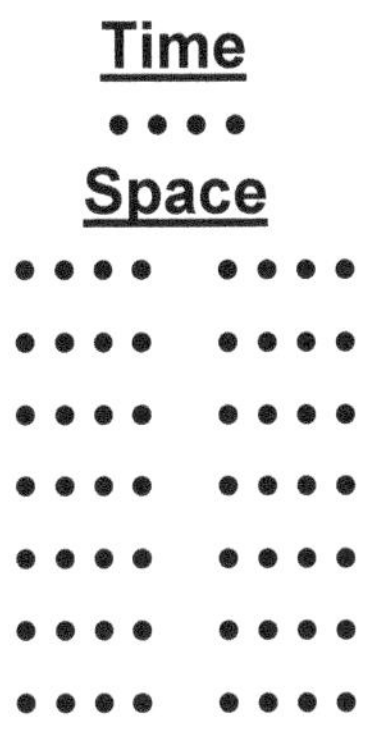

Time

• • • •

Space

• • • • • • • •

• • • • • • • •

• • • • • • • •

• • • • • • • •

• • • • • • • •

• • • • • • • •

• • • • • • • •

Figure 3.2. 8-dimensional subspace for determining internal symmetries and particle spectra.

We expect its 30 complex dimensions (60 real dimensions) will split into a 4-dimensional complex coordinate space[27] which will support Complex Lorentz transformations, and the remaining 26 complex coordinates will represent internal symmetry space. Internal symmetry space has an initial U(26) group before breakdown.

The mechanism for this symmetry breakdown may involve vacuum energy effects in the biquaternion universe.

The remaining four time dimensions (coordinates) shown in Fig. 2.1 will be shown to correspond to a U(2) group that maps between normal fermions and Dark fermions on a one-to-one basis. Experiment indicates this group's interactions must be extraordinarily weak.

In the next chapter we analyze the symmetries of the 26 complex dimensional subspace, and discuss the U(2) group in more detail.

---

[27] The 7 + 1 dimension biquaternion space yields a 3 + 1-dimensional complex coordinate space which then becomes the 3 + 1 dimensional real-valued space of our experience.

4. Internal Symmetries of the Quaternion Unified SuperStandard Theory (QUeST)

The coordinate space picture of QUeST described in chapter 3 enables us to simply find the internal symmetries and particle spectra of QUeST. They will turn out to be those of the Unified SuperStandard Theory.

4.1 QUeST Space

The QUeST coordinate subspace for the determination of internal symmetries is depicted in Fig. 4.1..It is based on the discussion of Fig. 3.2. The complex space-time 4-vector is separated from the internal symmetry coordinates in Fig. 4.1.

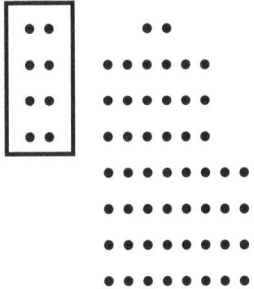

Figure 4.1. Symmetry determination subspace. Eight real-valued space-time dimensions (giving a complex 4-vector) are separated from the 52 real-valued dimensions for internal symmetries. The internal symmetry dimensions are split into subsets for each of the internal symmetries.

The internal symmetry coordinates number 52 real coordinate dimensions or 26 complex coordinate dimensions. These coordinates serve as coordinates of the fundamental representations of each of the factors of the total internal symmetry.

4.2 One Layer QUeST

The 26 complex coordinates form a U(26) internal symmetry space. This space undergoes a breakdown. The resulting factorized form of the internal symmetries[28] is given by

$$SU(2) \otimes U(1) \otimes SU(3) \otimes SU(2) \otimes U(1) \otimes SU(3) \otimes U(4)^4 \qquad (4.1)$$

The 7 + 1 dimension QUeST leads to a one layer form of the Unified SuperStandard Theory. In this theory there is one Layer group with a singlet representation.

When we generalize 7 + 1 QUeST to 32 complex quaternion dimensions we will have a QUeST with four layers that corresponds to the Unified SuperStandard Theory (UST) in 3 + 1 dimensions with the addition of a [U(2)]⁴ factor.

The one layer QUeST is represented by Fig. 4.2. The seemingly duplicate factors in eq. 4.1 of

$$SU(2) \otimes U(1) \otimes SU(3) \quad \otimes SU(2) \otimes U(1) \otimes SU(3) \qquad (4.2)$$

are for the "normal" and the Dark sectors respectively as shown in Fig. 4.2.[29] Fig. 4.2 reflects a map of the dimensions (coordinates) of Fig. 4.1 into representations of internal symmetry groups.

4.2.1 The U(2) Rotation Group Between Normal and Dark Sectors

The omitted four real-valued dimensions (coordinates) in Fig. 4.1 (compared to the complete set of dimensions in Fig. 2.1) are for the fundamental representation of a U(2) group. It appears that the only role this group could play is to transform between the Normal and Dark sectors. Thus we call it the *Dark Group*. It transforms each normal fermion to its Dark counterpart and *vice versa* (as well as rotations to fermions intermediate between normal and Dark.)

These observations are based on a realization that the Generation Group transforms among the four generations of the layer, and that the Layer group transforms among the four layers of fermions. Consequently the only remaining role the Dark U(2) group can play is between the normal and Dark fermion spectrums. Together these three

---

[28] Eq. 4.1 will be shown later to correspond to *one layer* of UST. UST has four layers, which are described later in this chapter.

[29] Simple counting of fundamental representation dimensions shows this to be true:: 2 + 3 + 2 + 3 = 10 respectively. The set of 10 complex coordinates support transformations to a factorized block-diagonal form. The 10 complex coordinates can be transformed into fundamental representations of the above factor group product.

types of groups "unite" the complete fermion spectrum. Fig. 4.3 shows the roles of the Generation, Layer and Dark groups for the fundamental fermion spectrum.

Experiment suggests the Dark U(2) group is necessarily broken with large mass vector bosons and/or with ultraweak interactions. Fig. 4.2 displays the one layer internal symmetry groups and complex space-time dimensions.

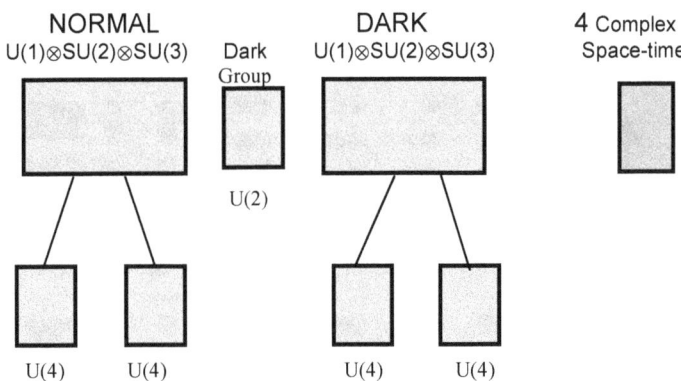

Figure 4.2. Schematic of the internal symmetry groups of eq. 6.1 plus 4 complex dimensions space-time. The two large blocks are each 5 dimension complex coordinate representations of SU(2)⊗U(1)⊗SU(3). The U(2) group supports transformations (rotations) between normal and Dark matter.

The Fermion Periodic Table

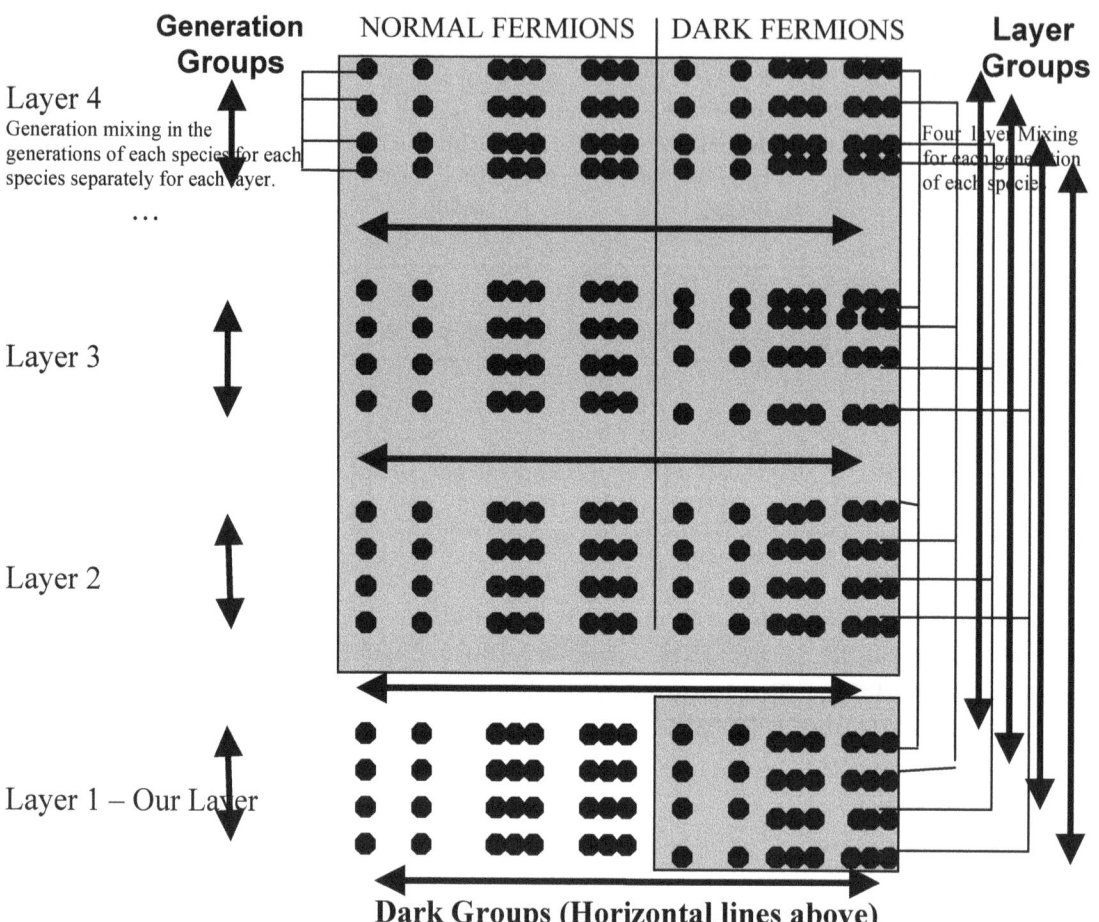

Figure 4.3. Fermion particle spectrum and partial example of pattern of mass mixing of the Generation, Layer, and Dark grroups. Unshaded parts are the known fermions with an additional, as yet not found, 4th generation shown. The lines on the left side (only shown for one layer) display the Generation mixing within each layer's species. The Generation mixing applies within each layer using a separate Generation group for each layer. The lines on the right side show Layer group mixing with the mixing amongst all four layers for each of the four generations individually. There are four Layer groups. The Dark groups

mixing between normal and Dark fermions are shown in the center as horizontal lines. There are 256 fundamental fermions counting quarks as triplets.

The Dark U(2) group transforms between normal particles and Dark particles. Thus a normal electron is mapped to the equivalent Dark electron, a normal neutrino is mapped to the equivalent Dark neutrino, a normal quark is mapped to the equivalent Dark quark, and so on for all fermions. Similarly normal vector bosons are mapped to Dark vector bosons and normal Higgs particles are mapped to Dark Higgs particles.

Examples of these mappings are:

$$\psi_d = U^{-1}\psi_n$$
$$A^{\mu}_d = U^{-1}A^{\mu}_n U$$
$$\psi_n = U\psi_d$$
$$A^{\mu}_n = U A^{\mu}_d U^{-1}$$

where d indicates Dark, n indicates normal, ψ is a fermion field, and A^{μ} is a vector boson field.

The combination of Generation, Layer and Dark groups interconnects, in principle, all parts of the fermion spectrum shown in Fig. 4.3. Thus "isolated" fermions do not exist in UST or QUeST.

4.2.2 Generation and Layer Groups

The lower U(4) groups in Fig. 4.2 are the Generation and Layer number groups. One pair of each number group is for each of the two U(1)⊗SU(2)⊗SU(3) factors above. Layer and Generation groups are part of the Unified SuperStandard Theory. See chapters 44 and 45 of Blaha (2020c) for detailed discussions.

4.3 The Breakdown of the Symmetry of the 20 Real Coordinate Group SU(10)

The 20 real coordinate subspace of Fig. 4.1 map to fundamental representations of

$$SU(2)\otimes U(1)\otimes SU(3)\otimes SU(2)\otimes U(1)\otimes SU(3)\otimes U(4)^4 \qquad (4.3)$$

which clearly can be transformed to a block diagonal form in a 26 dimensional group factors representation.[30]

---

[30] The separation of the 20 coordinate subspace into the subgroup factors representation can be implemented as group transformations and definitions using standard group theoretic methods. A more formal method for extracting the

However it is instructive to consider the generation of the component factors using a Lorentz boost-like framework. Separating the internal symmetry part[31] of Fig. 4.1 into six sets of Lorentz-like coordinates with a real energy and complex momenta we obtain Fig. 4.4.

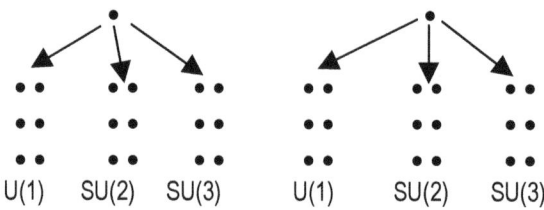

Figure 4.4. Each of the two real-valued time coordinates is linked to three complex spatial coordinates. Lorentz transformations applied to each 4-vector so constructed yields a factor of eq. 4.3. The factors so generated commute to give eq. 4.3 with commuting subgroup factors listed in the figure.

Each of the six sets of Lorentz-like coordinates have an embedded subgroup (listed in Fig. 4.4). These subgroups commute. Then taking each of the six different subgroups from the six subsets, that we can form, enables us to form the direct product in eq. 4.2. Thus the analogy between the Standard Model internal symmetries and Complex Lorentz subgroups is validated by the above construction.

4.4 Map to Unified SuperStandard Theory Groups

The map of the four complex coordinates to Complex space-time is direct. Complex Lorentz group transformations relate coordinate systems.

The complete set of internal symmetries (excepting the Dark group) for one layer is given by eq. 4.3. The map follows from the observations:

1. The factors of eq. 4.3 can be separated into separate factors for the normal and Dark sectors.

2. Since the set of normal and Dark fermions are split into four species[32] we can unambiguously associate the factors of eq. 4.3 with its factors.

---

subgroup content of representations uses a symmetric group analysis of $U(n)$ representation characters. See S. Blaha, J. Math. Phys. **10**, 2156 (1969) for a detailed discussion of this approach.

[31] Excluding the $U(4)^4$ and $U(2)$ parts.

[32] Appendix 5-B shows QUeST has four species (types) of fermions.

3. The SU(3) factors have <u>3</u> representations which we can associate with the up-quark and down-quark, normal and Dark species. Thus the SU(3) subgroups are normal and Dark Strong interaction subgroups.

4. The SU(2)⊗U(1) factors map to ElectroWeak interactions for the normal and Dark sectors.

5. The four U(4) factors map to the Generation and Layer number groups for the normal and for the Dark sectors.

Thus we have a map to the interactions of the *one layer* Unified SuperStandard Theory. The Dark U(2) group is added to the layer. Although we consider only one layer, the considerations apply to all four layers.

4.5 Four Layer QUeST Maps Exactly to UST Plus Dark U(2)^4

The QUeST figures above do not display the Internal Symmetry, Generation and Layer groups of all four layers of the Unified SuperStandard Theory. The groups of the layers are not the same. Each layer has its own set of groups;

$$SU(2) \otimes U(1) \otimes SU(3) \otimes SU(2) \otimes U(1) \otimes SU(3) \otimes U(4)^4 \qquad \text{See Fig. 4.2}$$

and

$$U(2) \qquad \text{Section 4.2.1}$$

The overall internal symmetry is the internal symmetry group of the Unified SuperStandard Theory augmented by U(2)^4 giving

$$[SU(2) \otimes U(1) \otimes SU(3) \otimes SU(2) \otimes U(1) \otimes SU(3) \otimes U(4)^4 \otimes U(2)]^4 \qquad (4.4)$$

The one layer theory is described by 7 + 1 dimension complex quaternion QUeST. The four layer theory is described by a 32 dimension complex quaternion QUeST. Thus it consists of four "copies" of the coordinates:

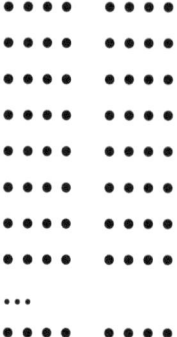

Figure 4.5. The 32 complex quaternion dimensions QUeST schematic.

which yield four duplicates of the internal symmetry schematic in Fig. 4.2. Note: there are four U(2) rotation groups – one for each layer,

4.6 Justification for a Four Layer QUeST

There is good reason for QUeST to have four layers embodied in 32 dimension complex quaternion space. If one considers the content of the layer displayed in Fig. 4.2 one sees a 4 dimension complex coordinates block for space-time. To create a 4 dimension complex quaternion coordinates space-time, one needs four layers of the form of Fig. 4.2. *The combination of the four 4-dimension complex coordinates is a complex quaternion dimensions space-time*. Thus the choice of four layer QUeST gives us a 4-dimension complex quaternion space-time AND enables QUeST to map directly to UST with its four layers if one limits the quaternion coordinates to the real-valued coordinates within them.[33]

We conclude four layer QUeST is needed to have a 4 dimension complex quaternion space-time.

---

[33] The Layer groups of UST enable mixing between the layers of fermions as shown in Fig. 4.3.

Appendix 4-A. Gauge Groups Based on Particle Numbers

In this Appendix[34] we show the origin of the Generation and Layer groups in particle number operators since they are not well known. See Blaha (2020c) for more details. Particle interactions followed directly in the Unified SuperStandard Theory by analogy with Complex General Relativity subgroups yielding

$$SU(2) \otimes U(1) \otimes SU(3) \otimes SU(2) \otimes U(1) \otimes SU(3) \qquad (4\text{-}A.1)$$

where the latter three factors are the Dark interactions.

They have a SU(10) covering group that contains this direct product of groups. The groups in eq. 4-A.1 are particle interaction groups in the Unified SuperStandard Theory.

Unlike other attempts to develop a formulation of the Standard Model (or generalizations) the Unified SuperStandard Theory was originally directly based on a theory foundation consisting of Complex General Relativity and Quantum Field Theory.

To those who might prefer to base a theory on real General Relativity we note that proofs in Quantum Field Theory *require* the Complex Lorentz Group.[35] Thus the Complex Lorentz group is unavoidable for a properly (and rigorously) formulated Quantum Field Theory. Since the formulation of the Complex Lorentz Group in flat space-time can only be as the limit of Complex General Relativity, the choice of a foundation of Complex General Relativity is required.

Since particles are countable, and thus have discrete particle numbers, Quantum Field Theory brings particle numbers, and particle number laws such as particle conservation laws, into consideration.

Blaha (2019e) and earlier books showed that Complex Lorentz boosts generate four types of fermion particles that we call *particle species*. We map these four species to charged leptons (such as electrons), neutral leptons (such as neutrinos), up-type quarks (such as the u quark), and down-type quarks (such as the d quark).

---

[34] This appendix is an extract from Blaha (2020c) for the readers convenience.
[35] Streater (2000).

4-A.1 Basis of the Generation Group

We define two particle number operators for normal up-quark particles and down-quark particles, B_{uq} and B_{dq}. Similarly we define two particle number operators for normal species "e" (electron) particles and species "v" particles, B_e and B_v. Similarly we define Dark matter equivalents:[36] B_{De}, B_{Dv}, B_{Duq}, and B_{Ddq}.

In the absence of interactions these fermion particle number operators are conserved. Each set are "diagonal" operators within a U(4) group. Thus we have a normal U(4) Generation Group and a Dark U(4) Generation group.

On this basis we find there are four generations of each species in the normal and in the Dark matter sectors. One generation of normal fermions with large masses has not as yet been found.

The gauge vector bosons of the Generation Group also have large masses. If the conservation of the fermion particle numbers is broken then we view it as a consequence of Generation Group symmetry breaking.

4-A.2 Basis of the Layer Group

The set of particle number operators can be further refined if we take account of the fourfold fermion generations. To further refine the set of particle number operators we temporarily neglect all interactions that would violate conservation laws for the set.

We therefore subdivide the above particle number set into four particle numbers per generation. For the ith generation we define

L_{ie} – The "e" species particle number for the ith generation
L_{iv} – The v species particle number for the ith generation
L_{iuq} – The up-quark species particle number for the ith generation
L_{idq} – The down-quark species particle number for the ith generation

L_{iDe} – The Dark "e" species particle number for the ith generation
L_{iDv} – The Dark v species particle number for the ith generation
L_{iDuq} – The Dark up-quark species particle number for the ith generation
L_{iDdq} – Dark down-quark species particle number for the ith generation

---

[36] By analogy, we assume that there are four species of Dark matter: charged Dark leptons, neutral Dark leptons, Dark up-type quarks, and Dark down-type quarks. Thus we are led to the Dark particle numbers: Dark Baryon Numbers, and Dark Lepton Numbers shown above.

for each generation i = 1, 2, 3, 4. Individual fermions have positive L_{ia} = +1 values and anti-fermions have negative L_{ia} = –1 values for species a = 1, 2, 3, 4 (with the three color subspecies of quarks treated as part of one species.)

At this point we have four particle number operators for each generation. We define a group framework for each set of particle numbers. The simplest way is to assume that each generation consists of four layers with the particles in each generation in a U(4) fundamental representation.[37] Then each generation has a U(4) Layer group with the generation's four number operators (above) as its diagonal operators. We call this group the Layer Group of the ith generation L_{ia}. With four generations we obtain four U(4) Layer groups for normal matter. In addition there are four U(4) Dark Layer groups.

The consequence of this expansion of particle numbers and groups is that the set of fermions increases fourfold. We now have four layers, with each having four generations, Experimentally, we know of three generations of fermions—the lowest generations of the lowest level. The remaining generation and three levels of fermions are of much higher mass and yet to be found.

See Blaha (2019g) and (2018e) for a detailed discussion of the Layer Groups. We note in passing that the symmetries of these number operators are badly broken. Yet the underlying group structure remains.

---

[37] See Fig. 2.3 for a depiction of the "splitting" of fermions: first into generations, then into layers.

5. QUeST Universes

QUeST is based on a 32 dimension complex quaternion space. It produces a 3 + 1 dimension complex quaternion space-time and the set of internal symmetries of the UST, to which $U(2)^4$ is added.

The similarities of QUeST and UST include the same four layer pattern; an identical normal and Dark fermion spectrum; and the same internal symmetries and vector bosons with QUeST adding the $U(2)^4$ Dark group vector bosons. See Appendix 5-A for the fermion and vector boson spectrums.

UST has additional detail that is consistent with the real-valued coordinates limit of QUeST. See Blaha (2020c).

***Upon restriction of QUeST to real-valued coordinates,
QUeST Becomes the Unified SuperStandard Theory.***

5.1 Are All Universes the Same?

Assuming that there are other universes beyond our own universe, the question of the Physics of other universes arises. If all universes are 32 dimension complex quaternion universes based on QUeST, then it appears their gross features would be similar. As we showed in Blaha (2020c), and in earlier books, coupling constants appear to be determined by vacuum polarization eigenvalue conditions.

The space-time, Internal Symmetries, fermion spectrum, and the vector boson spectrum are determined in QUeST. Thus the only remaining physical parameters needed are particle masses, and particle mixing. These parameters appear to be handled by vacuum considerations, which should be similar in all sufficiently large universes.

We conclude that all universes are the same based on these considerations. When we consider a complex octonion space, that we call the Megaverse, we will see that it is possible that 6 different varieties of universe may exist within the Megaverse.

Appendix 5-A. Unified SuperStandard Theory (UST) in 3 + 1 Real Dimensions

This Appendix presents a summary of the main consequences of our Unified SuperStandard Theory in 3 + 1 real dimensions. We see that it is the limit of the Quaternion Unified SuperStandard Theory (QUeST) in 32 Complex Quaternion Dimensions when it is restricted to 3 + 1 real-valued space-time coordinates.

The internal symmetries (interactions) of the theory, the vector boson spectrum, and the fermion spectrum are presented. QUeST predicts all these results in 32 dimension complex quaternion space. ***Thus QUeST is the ground of the Unified SuperStandard Theory in 3 + 1 real dimensions. UST was derived previously based on real-valued 3 + 1 dimensional coordinates, Complex General Relativity and Quantum Field Theory.***

5-A.1 Axioms and Derivation of the Unified SuperStandard Theory in 3 + 1 real dimensions

The axioms and derivation of the Unified SuperStandard Theory are presented in Blaha (2020c).

5-A.2 Unified SuperStandard Theory Interactions

The vector boson symmetries[38] in the Unified SuperStandard Theory were found to be

$$[U(1) \otimes SU(2) \otimes SU(3) \otimes U(1) \otimes SU(2) \otimes SU(3) \otimes U(4)^4]^4 \otimes U(4) \qquad (5\text{-A}.1)$$

QUeST causes us to add a new $U(2)^4$ Dark groups factor. The symmetries are of similar form in each of the four layers. In addition the symmetries of the "normal" sector and the Dark sector are duplicates.

QUeST has the same internal symmetries and adds an additional U(2) symmetry for each of the four layers. Thus $U(2)^4$. This symmetry maps each normal fermion to its corresponding Dark equivalent, and vice versa, layer by layer. We discuss this symmetry in more detail later.

The symmetries of the normal sector are

---

[38] Most of these symmetries are broken symmetries.

$$[U(1) \otimes SU(2) \otimes SU(3) \otimes U(4)^2]^4 \otimes U(4) \qquad (5\text{-}A.2)$$

where the lone U(4) is that of the Species group described later.

The symmetries of the Dark sector are similar (but different0

$$[U(1) \otimes SU(2) \otimes SU(3) \otimes U(4)^2]^4 \otimes U(4) \qquad (5\text{-}A.3)$$

where the lone U(4) is again that of the Species group.

Each of the four layers of fermions experiences its own set of interactions. They must be different since inter-layer interactions are not seen. Similarly the normal and Dark sectors of each layer must also have different sets of interactions. The U(2) groups added by QUeST must also yield ultra-weak interactions since interactions between normal and Dark matter are not seen.

The Layer groups act to interconnect the four layers for the normal and Dark layers separately. The U(2) groups serve to connect the normal and Dark sectors (in principle). Thus the symmetries result in an interconnected spectrum of fermions.

5-A.3 Unified SuperStandard Theory Vector Bosons

Fig. 5-A.1 displays the Unified SuperStandard vector bosons except for the Species group which all particles experience, and except for the U(2) groups added by QUeST.

The "Normal" Vector Boson Spectrum

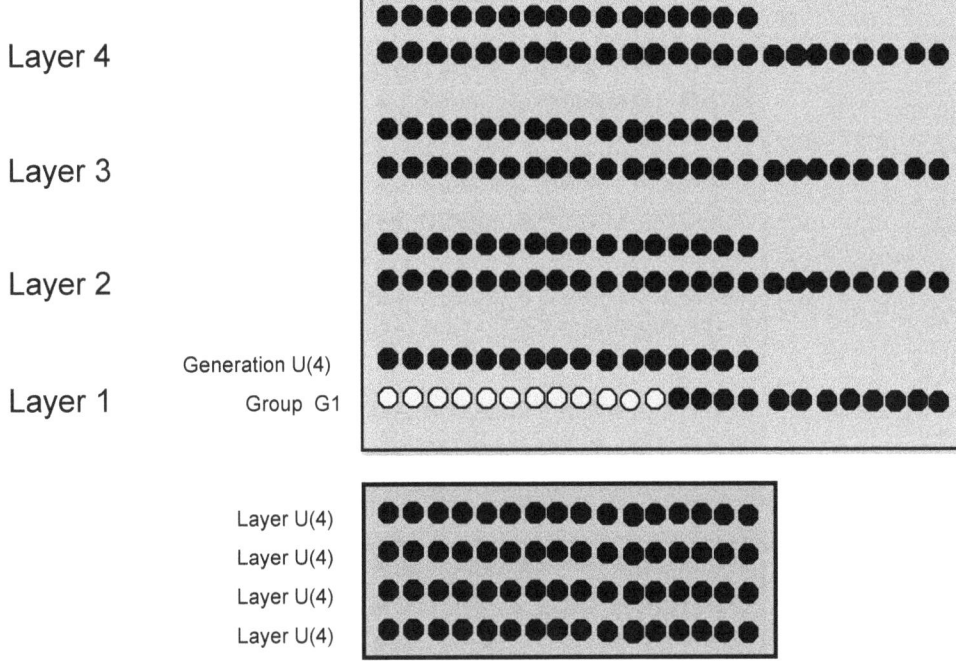

Layer 4

Layer 3

Layer 2

Generation U(4)

Layer 1   Group G1

Layer U(4)
Layer U(4)
Layer U(4)
Layer U(4)

Figure 5-A.1. The vector bosons. Each circle represents a group generator. The known vector bosons are in the lowest row with a white interior. Yet to be found vector bosons are solid black. The Layer groups straddle all four layers. G1 is SU(3)⊗SU(2)⊗U(1). The list of groups for the higher three levels is the same as those of the first layer. There are 224 normal vector bosons not counting Species and Dark U(2) groups.

The "Dark" Vector Boson Spectrum

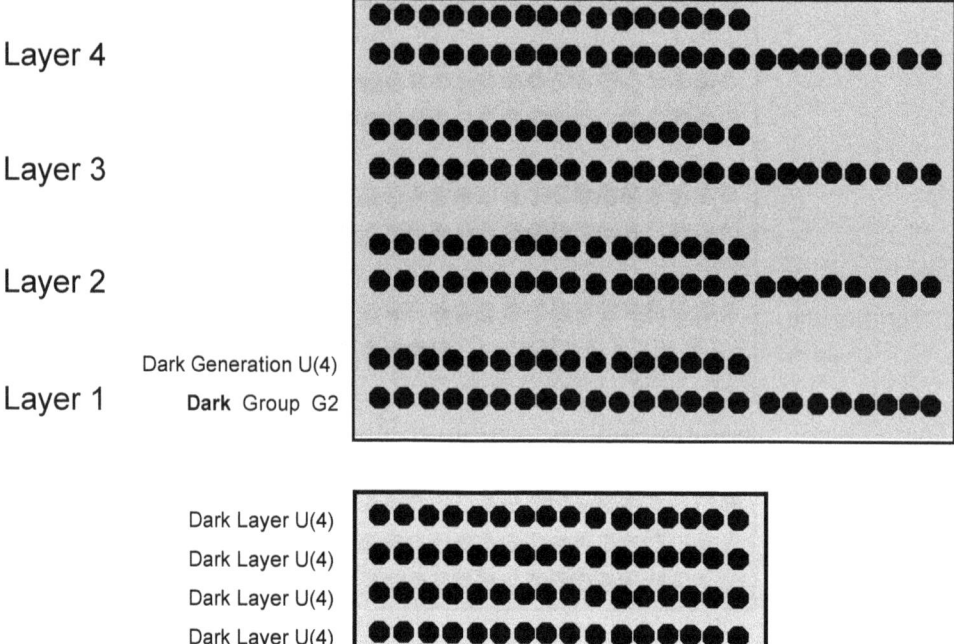

Figure 5-A.2. The vector bosons. Each circle represents a group generator. The known vector bosons are in the lowest row with a white interior. Yet to be found vector bosons are solid black. The Layer groups are distributed by layer symbolically although they each straddle all four layers. G2 is $SU_D(2) \otimes U_D(1) \otimes SU_D(3)$. The list of groups for the higher three levels is the same as those of the first layer. There are 224 Dark vector bosons not counting Species and Dark U(2) groups.

5-A.4 Fermion Mass Spectrum

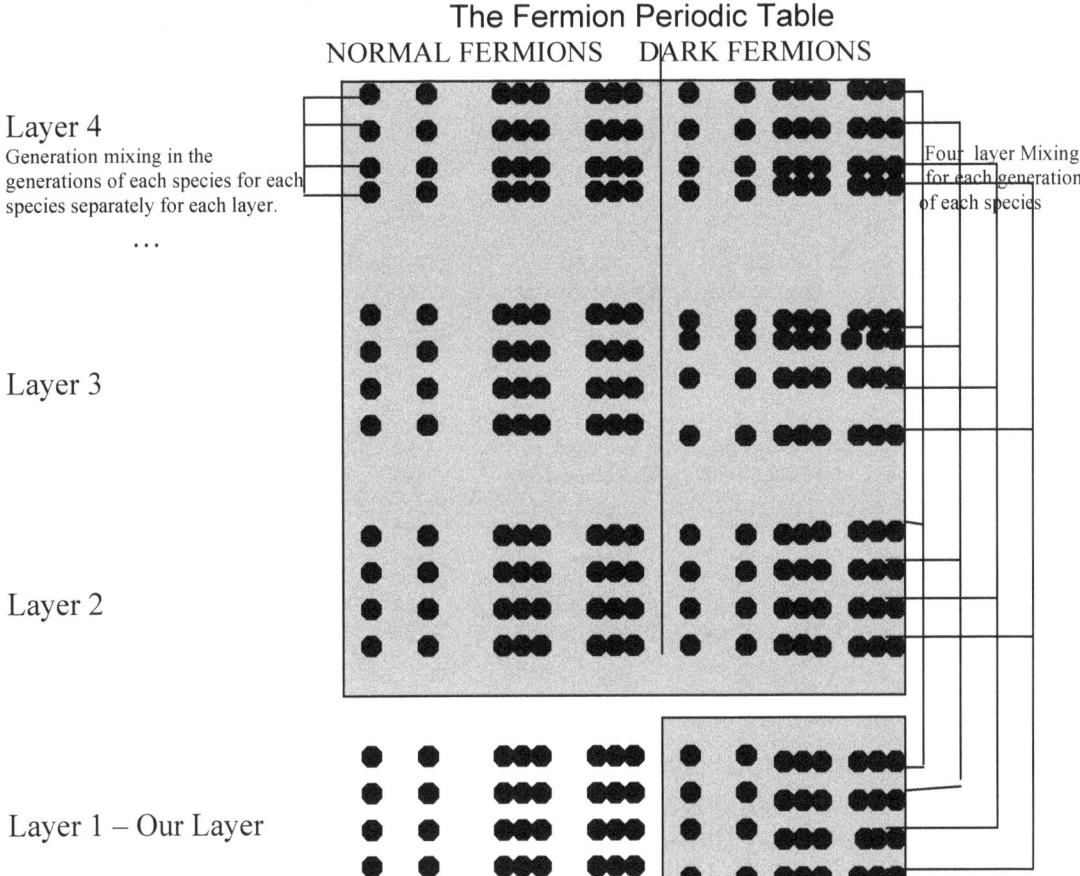

Figure 5-A.3. Fermion particle spectrum and partial example of pattern of mass mixing of the Generation group and of the Layer group. Unshaded parts are the known fermions with an additional, as yet not found, 4th generation. The lines on the left side (only shown for one layer) display the Generation mixing within each layer's species. The Generation mixing applies within each layer using a separate Generation group for each layer. The lines on the right side show Layer group mixing with the mixing amongst all four layers for each of the four generations individually. There are four Layer groups. There are 256 fundamental fermions. QUeST has the same fermion spectrum.

5-A.5 Groups and Fermion Splittings

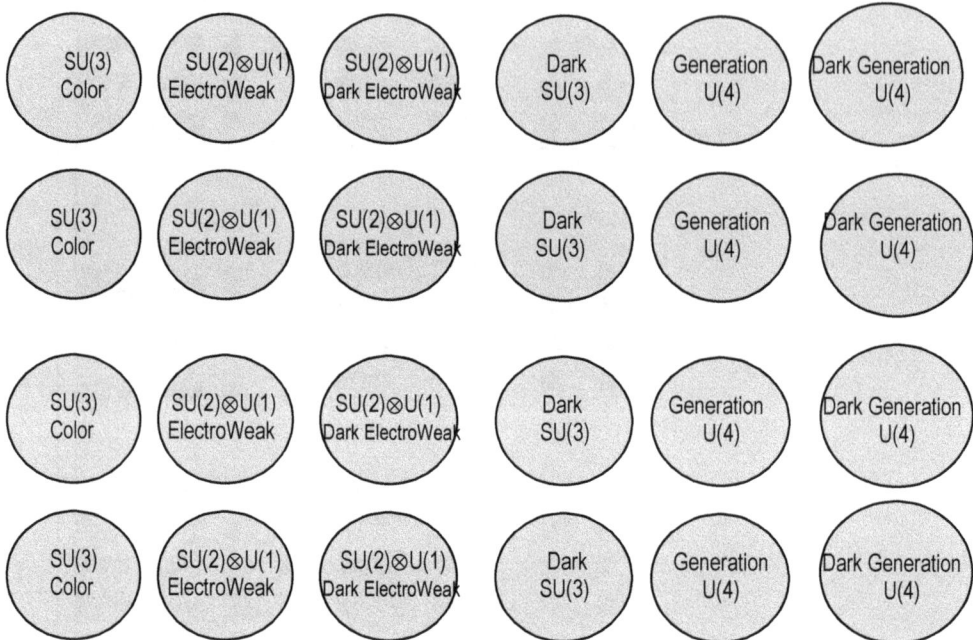

Figure 5-A.4. The set of four layers of internal symmetry groups corresponding to four generations in four layers of spin ½ fermions and the four layers of vector bosons. In addition there are the Normal and Dark Layer groups, the Species group, and the Dark U(2) groups, which are *not* displayed.

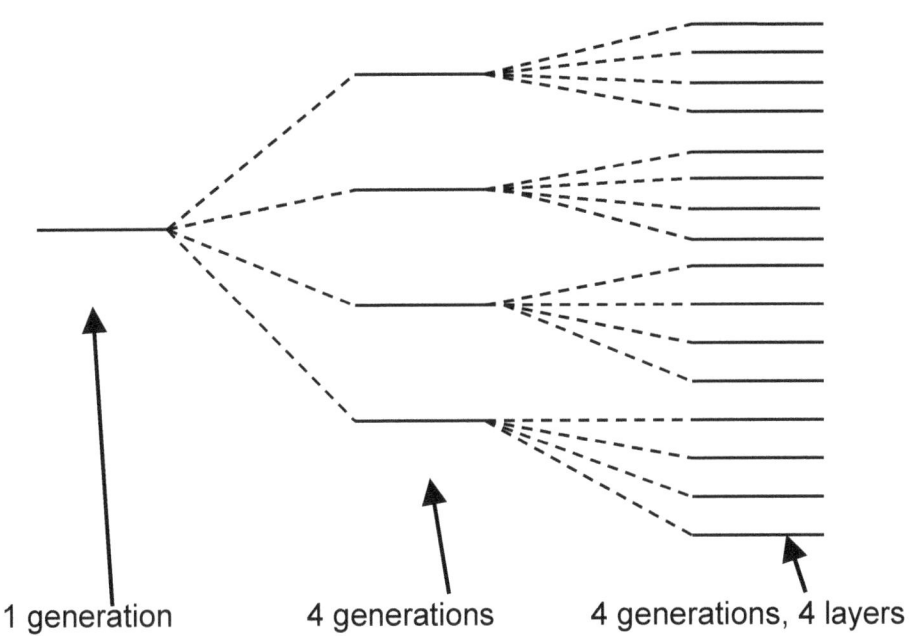

1 generation   4 generations   4 generations, 4 layers

Figure 5-A.5 The "splitting" of a single generation fermion into four generations and then into four layers.

Appendix 5-B. Four QUeST Fermion Species

QUeST is defined in a 3 + 1 dimensional complex quaternion space. These coordinates support a Complex Quaternion Lorentz Group just as the United SuperStandard theory. There are four types of boosts in the 3 + 1 Complex Quaternion space-time. Correspondingly, there are four species of QUeST fundamental fermions: "charged" lepton species, neutral lepton species, up-type quark species, and down-type quark species. The nature of each species is the same as in the Unified SuperStandard Theory.

The boosts are

1. Boosts from a rest frame to a rest frame with a relative velocity less than c.
2. Boosts from a rest frame to a rest frame with a relative velocity greater than c.
3. Boosts from a rest frame to a rest frame with a relative complex velocity less than c.
4. Boosts from a rest frame to a rest frame with a relative complex velocity greater than c.

The key relation for boosts in complex quaternion space is

$$e^x = e^a \left(\cos(\|\mathbf{v}\|) + \mathbf{v}/\|\mathbf{v}\| \, \sin(\|\mathbf{v}\|)\right)s \qquad (5\text{-}B.1)$$

using definitions in Appendix 2-A, in analogy with the similar complex-valued identity used in Blaha (2020c).

6. Bioctonion Megaverse

Earlier we developed a biquaternion theory called QUeST that could be used to derive the internal symmetry group structure, and the fundamental fermion and vector boson spectrums. It led to the Unified SuperStandard Theory (UST) and accounted for the close similarity between the internal symmetries of the Standard Model sector and the subgroups of the Lorentz group. (They both exhibit U(1), SU(2) and SU(3) symmetries.)

6.1 MOST

If we assume the existence of a Megaverse[39] containing our universe, and other universes, then we can define a *Megaverse Octonion SuperStandard Theory* (MOST) that becomes a more general basis for QUeST and the Unified SuperStandard Theory.

In this chapter we define a bioctonion space and use it to define a Megaverse basis for MOST. MOST develops a more robust set of internal symmetries and fundamental particles. It creates a new view of Dark matter that appears to help explicate the lack of interactions between normal matter and Dark matter.

Remarkably MOST, when "restricted" to our universe, yields QUeST.

6.2 Motivation and Procedure

Our goal again is to create a larger dimension space within which a universe based on QUeST can exist, and where we can ultimately derive our space-time and the Unified SuperStandard Theory. Again we use the similarity of Lorentz subgroups and Standard Model internal symmetry groups in our development.

Again there are two possible procedures to follow in developing the deeper basis:

1. One can develop the Quantum Mechanics, Quantum Field Theory, … in an octonion space and then extract the dynamics, fermion spectrum, gauge fields, and so on of our familiar space-time.

2. One can define an octonion space and then use its coordinates to directly extract the space-time, internal symmetries, fermion spectrum, gauge field spectrum and dynamics.

---

[39] The Megaverse is described in some detail in Blaha (2017c), (2017f), and (2018e) together with evidence for its existence.

Again we have chosen the latter approach.

In developing the deeper space, upon which we build, we will take guidance from the derivation of the Unified SuperStandard Theory. This theory assumes a complex 4-dimensional space-time upon which Complex General Relativity is constructed. It then proceeds to complex flat space-time and Complex Relativity. Similarly, QUeST is based on a complex coordinate system: complex quaternion space.

After defining features of Complex Lorentz transformations the Unified SuperStandard Theory used Lorentz boosts to derive the Dirac forms of the four fermion species. The boosts were required to boost a fermion from a rest state to a state with a real-valued energy, and real-valued or complex-valued 3-momenta. Thus a real time –complex-valued spatial part is required for the proper definition of species.

Similarly we define a complex octonion (bioctonion) space, and use the real octonion time, and the complex octonion spatial part to define the subspace for internal symmetries.

As in the previous quaternion case (section 4.3), we will show that the MOST internal symmetry subgroups map to Standard Model internal symmetry subgroups.

6.3 Definition of Bioctonion Space

Following the above stated procedure we define a bioctonion[40] (complex octonion) space with *one* "time" biquaternion and *seven* "spatial" bioctonions as a generalization of the 3 + 1 space-time of our experience. The choice of 8 bioctonion dimensions seemed natural but was not required by a principle. It does lead to a larger set of internal symmetries with *one layer* QUeST symmetries as a subset.[41] We will use the psiphi symbol • to represent each of the bioctonion space coordinates in Fig. 6.1.

We have chosen a complex 8-dimensional bioctonion space-time as the one MOST layer Megaverse space-time. There are 128 real coordinates in the bioctonion Megaverse space from which complex 8-dimensional real space-time (4-dimensional complex space-time) is extracted. Our universe's 4-dimensional complex space-time is a subspace-time.

Fig 6.1 symbolically depicts the space with a psiphi • for each real-valued coordinate. *Again we treat the bioctonion space as a higher dimensional space and do not use details of octonion algebra in our development.*

---

[40] We use bioctonion synonymously with complex octonion in this and subsequent chapters.
[41] **We start the complex octonion discussion by developing a one layer MOST. Then we develop a four layer MOST which contains four layer QUeST-based universes.**

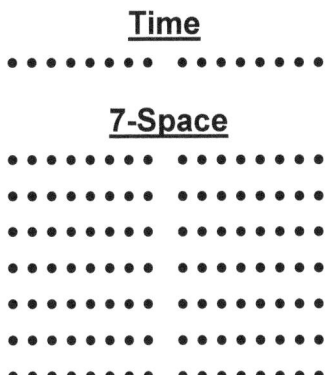

Figure 6.1. Eight-Dimensional (7 + 1) bioctonion space with coordinates represented by • 's. This bioctonion space has 128 real dimensions (64 complex dimensions.).

6.4 Bioctonion Lorentz Group

Our definition of time and space bioctonion coordinates purposefully resembles those of our real space-time. One might ask why there should be a Lorentz-like group for bioctonion space.

The only apparent reason is the need for a special speed c whose existence enables one to boost from the rest frame of a mass m particle with rest energy m to a moving frame of greater energy.[42] Without c the group of the above coordinates would presumably be the bioctonion U(8) group. U(8) transformations preserve the norm of a state so that a "boost-like" transformation does not exist.

The bioctonion Lorentz transformations do have a unique speed c (the speed of light) and specify a unique rest frame for any particle—both sublight particles and tachyon particles.[43] . Thus we select the bioctonion Lorentz group for bioctonion space.

Flat space bioctonion Special Relativity generalizes to a bioctonion General Relativity which may be constructed directly (mindful of octonion non-commutativity).

Flat space bioctonion Lorentz group transformations have constant bioctonion matrix elements that are analogous to those of the Lorenz group.

---

[42] A similar consideration applies to tachyons.
[43] Tachyons can be transformed by a Complex Lorentz transformation to and from a rest frame. See Blaha (2018f).

7. Symmetries of the Megaverse Octonion SuperStandard Theory (MOST)

The coordinate space-time picture of the Megaverse described in chapter 6 enables us to simply find the internal symmetries and particle spectra of MOST. They will turn out to be a superset of those of the Unified SuperStandard Theory.

7.1 Extracting the Symmetries and Particle Spectra

As stated earlier in chapter 4 for the quaternion case, we will directly describe the symmetry structure implied by the form of the bioctonion (complex octonion) coordinate system.

The Unified SuperStandard Theory developed the group structure, from which the particle species were derived, from a subset of Lorentz boost transformations. Complex boosts mapped a system at rest to a system in motion with a real energy and complex 3-momenta in general. Bioctonion Lorentz boosts play a similar role.

The 4-dimensional representation of the Unified SuperStandard Theory complex coordinates subset is given in Fig. 7.1.

Time
•
Space
••
••
••

Figure 7.1. Four-Dimensional space for Unified SuperStandard derivation of particle spectra with coordinates represented by psiphi • 's.

Following the same line of thought, which is described in chapter 4, we now specify a bioctonion subspace restricted to that of Fig. 7.2 to define the relevant set of coordinates for determining particle symmetries and spectra.

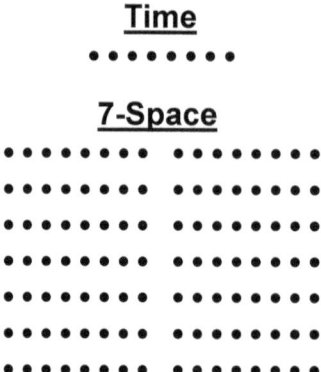

Figure 7.2. 8 complex dimension bioctonion subspace for symmetries and particle spectra.

The subspace's 60 complex dimensions (coordinates) will split into an 8-dimensional complex coordinate space which supports 8-dimensional Complex Lorentz transformations, and a 52 complex dimension (coordinates) internal symmetry space. The neglected eight dimensions support a Dark U(4) rotations group[44] transforming between normal and Dark sectors. (There are four sectors in MOST necessitating the U(4) rotations that transform among them.)

The mechanism for this symmetry breakdown may be due to vacuum energy effects in the bioctonion Megaverse.

The MOST Megaverse subspace for the determination of the Internal Symmetries is depicted in Fig. 7.3. It is based on the discussion above.

---

[44] This Dark group is the analogue of the complex quaternion Dark U(2) group.

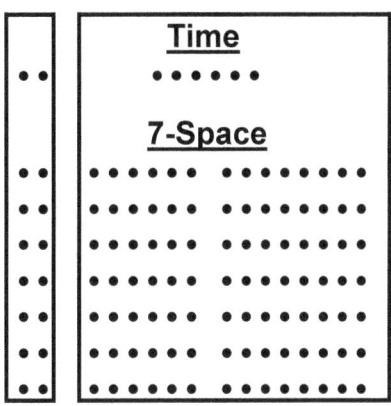

Figure 7.3. Complex 7 + 1 space-time (left box) and Internal Symmetry determination subspace (right box). The 16 real space-time coordinates are separated from 104 real coordinates for internal symmetries.

The internal symmetry coordinates above number 104 real coordinates or 52 complex coordinates. These coordinates serve as the coordinates of the fundamental representations of each of the factors of

$$[SU(2)\otimes U(1)\otimes SU(3)\otimes SU(2)\otimes U(1)\otimes SU(3)]^2\otimes U(4)^8 \qquad (7.1)$$

The factorized internal symmetry emerges from another breakdown(s) which corresponds to the subgroup structure of the Lorentz group. Eq. 7.1 evidently follows from the structure of the bioctonion Lorentz transformations.

The U(4) Generation and Layer groups are represented in Fig. 7.1. We depict the pattern of symmetry implied by Fig. 7.1 and eq. 7.1 in Fig. 7.4 and 7.5 below.

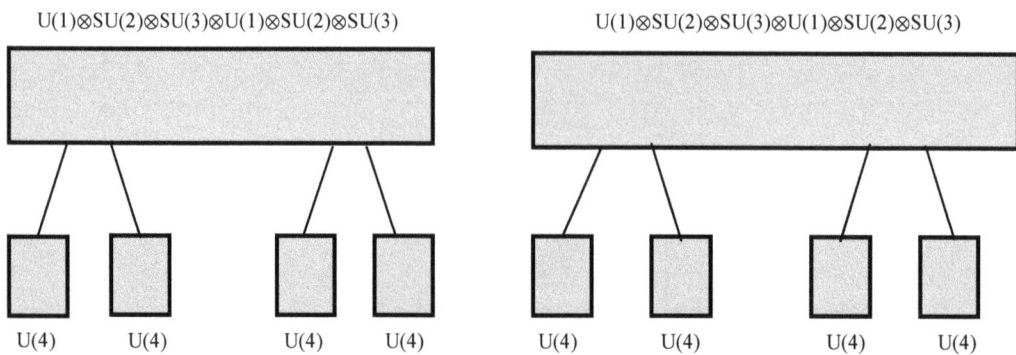

Figure 7.4. Schematic of the internal symmetry groups' coordinates of Fig. 7.1. The two "large" blocks are each sets of 20 real-valued coordinates furnishing representations of the indicated groups. The lower U(4) groups are the Generation and Layer number groups. The Dark U(4) group is not shown. The total number of real-valued coordinates is 104.

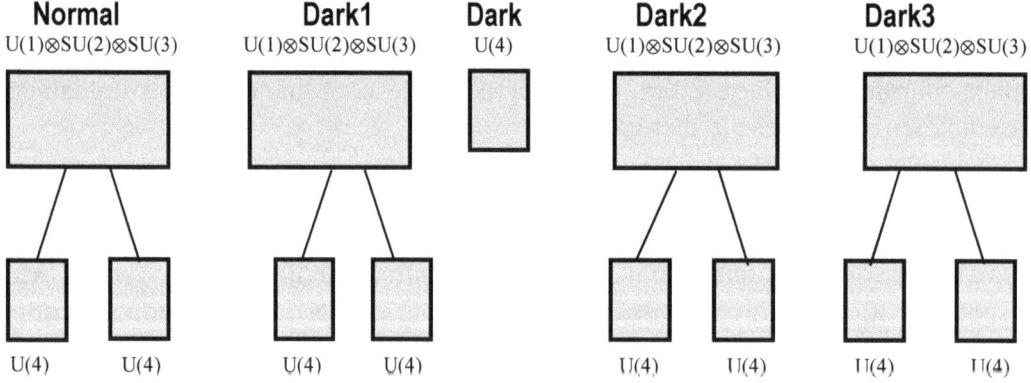

Figure 7.5. Schematic of the internal symmetry groups of eq. 7.1 with the Dark U(4) group added. These are the internal symmetry groups of one layer MOST. The lower U(4) groups are the Generation and Layer number groups. One pair of each number group is for each of the four U(1)⊗SU(2)⊗SU(3) factors above. The result is the total Internal Symmetry group of one layer of an generalization of QUeST and the Unified SuperStandard Theory.

Each U(1)⊗SU(2)⊗SU(3)⊗U(1)⊗SU(2)⊗SU(3) block in Fig. 7.4 has a 10 complex coordinates (20 real-valued coordinates) representation. The blocks are subdivided in Fig. 7.5 into sets of 10 real-valued coordinates supporting representations of U(1)⊗SU(2)⊗SU(3). We assign the first block to contain the representations of the known parts of the Standard Model. There are three Dark blocks. The internal symmetry groups of each part are listed in Fig. 7.5.

The factorization of each of the four blocks is accomplished by following the procedure given in section 4.3 for each block.[45]

7.2 One Layer MOST

The above section specifies the *one layer* MOST. The symmetries of the three other layers are the same but their groups are individual to each layer. The groups of each layer are each flagged with a different index.

The overall one layer MOST internal symmetry is specified by Fig. 7.5.

$$[SU(2)\otimes U(1)\otimes SU(3)]^4 \otimes U(4)^9 \qquad (7.2)$$

There is also a space-time sector with 8 (= 7 + 1) complex coordinates.

Comparing Figs.4.2 and 7.5 we see that a one layer QUeST has a "Normal" part and one Dark part while the MOST adds two more Dark parts.

Thus the internal symmetry groups of one Layer MOST are:

"Normal" Gauge Groups
SU(3)⊗SU(2)⊗U(1)
Generation Group U(4)
Layer Group U(4)
Dark1 Gauge Groups
SU(3)⊗SU(2)⊗U(1)
Generation Group U(4)
Layer Group U(4)

Dark2 Gauge Groups
SU(3)⊗SU(2)⊗U(1)
Generation Group U(4)
Layer Group U(4)

---

[45] The separation of the 20 coordinate subspace into the subgroup factors representation can be implemented as group transformations and definitions using standard group theoretic methods. A more formal method for extracting the subgroup content of representations uses a symmetric group analysis of U(n) representation characters. See S. Blaha, J. Math. Phys. **10**, 2156 (1969) for a detailed discussion of this approach.

<u>Dark3 Gauge Groups</u>
SU(3)⊗SU(2)⊗U(1)
Generation Group U(4)
Layer Group U(4)

PLUS

A Dark U(4) group that rotates among the four normal and Dark sectors

Figure 7.6. One layer MOST vector bosons list from eq. 7.2. The four layer MOST quadruples the above list: with one distinct set for each layer. In one layer the total number of vector bosons of the above list is 192 for one layer. Thus four layers yield a total count of 768 vector bosons in MOST (not counting the Species group which comes from General Relativity). We require each level has a separate Dark U(4) rotation group.

7.3 Four Layer MOST

The four layer MOST is described by a 32 dimension complex octonion space. Thus it consists of four "copies" of the coordinates:

Figure 7.7. The 32 dimension MOST schematic. Four layer MOST has 512 real-valued dimensions.

which yield four duplicates of the internal symmetry schematic in Fig. 7.5,

and an 8 complex quaternion space-time consisting of 7 + 1 complex-valued quaternion coordinates.

The sum total of real-valued dimensions is 512 as is the sum of the dimensions of the above parts constructed from the dimensions.

In short we obtain a four layer MOST

which may contain QUeST universes.

7.4 Justification for a Four Layer MOST

There is good reason for MOST to have four layers embodied in 32 dimension complex quaternion space. If one considers the content of a layer displayed in Fig. 7.3 one sees a 8-dimension complex coordinates block for space-time. To create an 8 dimension complex *quaternion* coordinates space-time, one needs four layers of the form of Fig. 7.3. *The combination of four 8-dimension complex coordinates is an 8 dimension complex quaternion space-time.*

Thus the choice of four layer MOST gives us an 8-dimension complex quaternion space-time that can contain 4-dimension complex quaternion QUeST universes. *We conclude four layer MOST is needed to have an 8-dimension complex quaternion space-time.*

The thirty-two dimension complex octonion space contains an 8-dimension complex quaternion space-time and the four layers of Internal Symmetry groups shown in Fig. 7.5.

7.5 Fermion and Gauge Vector Boson Spectrums

The fermion and vector boson spectrums that emerge in MOST are those of an "enlarged" QUeST and Unified SuperStandard Theory. They are displayed below. MOST has an additional two Dark sectors beyond QUeST and the Unified SuperStandard Theory.

Vector Bosons

From Fig. 7.6 we find MOST has 192 vector bosons in one layer. Thus four layer MOST has a total count of 768 MOST vector bosons. There are two additional Dark vector boson sectors beyond QUeST and the Unified SuperStandard Theory.

Fermions

There are 512 fundamental fermions in MOST, which includes two additional Dark fermion sectors. Fig. 7.8 shows the MOST fermion spectrum.

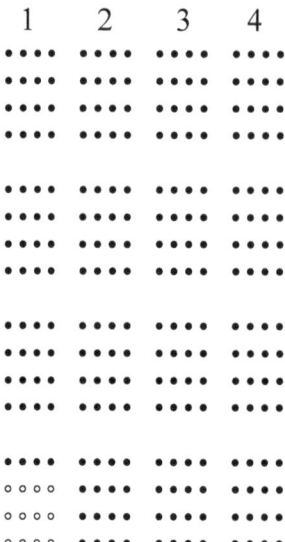

Figure 7.8. Schematic spectrum of the fermions of 4 layer MOST. Each fermion is represented by a •. Quark triplets are represented by a single •. Four sets of four species in four generations which are in turn in 4 layers. Open symbols ○ represent known fermions. There are 512 fundamental fermions taking account of quark triplets. Note the Layer groups determine the layers in UST. **They require 4 layers of 8 complex octonions in Megaverse space leading to the 32 dimension complex octonion space.**

8. The MOST Megaverse

The MOST Megaverse *space-time* has 8 complex quaternion dimensions (64 real-valued dimensions counting the 8 dimensions within each of the eight quaternion coordinates). It has a larger set of Internal Symmetries and particles than a QUeST universe. Four layer QUeST universes can exist in a four layer MOST Megaverse.

MOST has four U(1)⊗SU(2)⊗SU(3) parts (Fig. 7.5) with a different set of U(1)⊗SU(2)⊗SU(3) gauge vector bosons in each part. It also has four blocks of fermions (shown in Fig. 7.8).

8.1 QUeST Universes Within MOST

Given the four parts of MOST Internal Symmetries the question arises as to how a QUeST universe, which has only two parts, is constructed. The four parts have the same groups, and the same associated fermion and vector boson particles. The author has shown that the coupling constants of the groups are determined by vacuum polarization effects. (See Blaha (2020c).) The remaining features: particle masses and symmetry breaking, are determined by vacuum effects, which suggests that they may be the same for all four parts of MOST.[46]

If we construct a universe (conceptually) there are 6 possible combinations of the four parts, from which two parts are selected. If all parts are the same, as the previous paragraph suggests, then all combinations lead to the same QUeST universe. If the parts differ then there would be up to 6 variants of QUeST universes. Our universe then becomes one possibility.

---

[46] In the Megaverse, the Megaverse vacuum determines these quantities. In a universe, its vacuum determines these quantities.

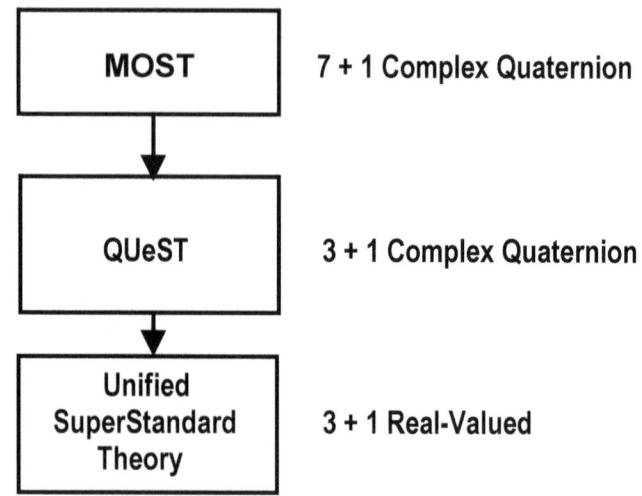

Figure 8.1. The hierarchy of Unified SuperStandard Theories.

8.2 MOST Spinors

Note the 7 + 1 Complex Octonion Megaverse's space-time has 16 component Dirac fermion spinors.

8.3 MOST Fermion Species

The fermion species in MOST number four despite the complex eight-dimensional nature of the space-time extracted from complex octonion space. The eight complex space-time coordinates support a 7+1 Complex Quaternion Lorentz Group.

There are four types of boosts in this 7+1 complex quaternion space-time. They correspond to the types of boosts in UST. They boost a particle rest state to a state of motion with a real energy p^0 (the magnitude of the quaternion-valued energy), and a real quaternion valued or complex quaternion valued momentum (spatial coordinates) whose magnitude $|\mathbf{p}|$ is a real-valued or complex-valued quantity. The fermions that can be generated by 7 + 1 Complex Quaternion Lorentz Group boosts[47] are:

1. A boost from rest to a frame with real-valued energy and momentum with $p^{02} - \mathbf{p}^2 > 0$. A "normal charged lepton-like" fermion.

---

[47] The boosts rely on eq. 2-A.4.

2. A boost from rest to a frame with real-valued energy and momentum with $p^{02} - \mathbf{p}^2 < 0$. A "tachyonic neutral lepton-like" fermion.

3. A boost from rest to a frame with real-valued energy and a complex-valued spatial momentum with $p^{02} - \mathbf{p}^2 > 0$. An "up-type quark-like" fermion.

4. A boost from rest to a frame with real-valued energy and momentum with $p^{02} - \mathbf{p}^2 < 0$. A "tachyonic down-type quark-like" fermion.

where p^0 is the energy magnitude and $\mathbf{p}$ is a spatial 7-vector magnitude. The particle states generated by 1 and 2 are "conventional" 8-dimensional analogues of the 4-dimensional case.

8.4 Octoquarks

The quark-like cases have an 8-dimensional aspect that prompts us to call them *octoquarks*. In 4-dimensional complex space-time a quark momentum is set by the term

$$\omega \mathbf{w} = \mathbf{u}_r \omega_r + i\mathbf{u}_i \omega_i \qquad (8.1)$$

as shown in Blaha (2020c). We now limit the 8-dimensional complex quaternion space-time to a complex space-time. In 8-dimensional complex space-time the 8-momentum, the *hyper-momentum*,[48] is set by

$$\omega \mathbf{w} = \mathbf{u}_r \omega_r + i\mathbf{u}_r \omega_1 + j\mathbf{u}_2 \omega_2 + k\mathbf{u}_3 \omega_3 + q\mathbf{u}_4 \omega_4 + r\mathbf{u}_5 \omega_5 + s\mathbf{u}_6 \omega_6 \qquad (8.2)$$

where the $\mathbf{u}_i$ are 8-vectors satisfying

$$\mathbf{u}_i \cdot \mathbf{u}_j = \delta_{ij} \qquad (8.3)$$

and i, j, k, q, r, and s are fundamental quaternion-like units.

Note that a fermion spin s satisfies the 8-vector inner product

$$s \cdot p = 0 \qquad (8.4)$$

or

$$\mathbf{s} \cdot \mathbf{p} = 0 \qquad (8.5)$$

in a manner similar to 4-space-time.

---

[48] As opposed to complex momentum in 4-space-time.

Thus the seven 7-vectors in 8-space-time form an orthonormal set and define quarks with 7-momentum. We call them *octoquarks* since the space-time is 8-dimensional.

Normal and tachyonic particles are distinguished in complex 8-dimensional space by the sign of $p^{02} - \mathbf{p}^2$ in a manner similar to 4-space-time. MOST fermions occur in four species just like QUeST fermions and Unified SuperStandard Theory fermions.

There are four species of fundamental fermions in 4-space-time and 8-space-time: "charged" lepton species, neutral lepton species, up-type octoquark species, and down-type octoquark species. The nature of each species (normal or tachyon, and real-valued or complex spatial momentum) is the same as in the Unified SuperStandard Theory.

9. Particle-Dimension Duality

In 32 dimension complex quaternion space there are 256 real dimensions and 256 fundamenatl fermions. In 32 dimension complex octonion space there are 512 real dimensions and 512 fundamenatl fermions. The equality of these parameters raises the possibility that the number of space dimensions equals the number of fundamental fermions in some deep sense. This possibility is enhanced because the number of particles depends on the singlets nature of leptons and the triplet nature of quarks.

When one considers the measurement of distance in a space one sees the measuring process requires the use of particles. Without particles one cannot distinguish space-time points from each other. And the distance between space-time points is determined either directly or indirectly by bosons such as light transmitted (and possibly reflected) between fermion "clumps.'

Without particles distance, space-time is meaningless. Without space-time particles will "clump' and dynamics is impossible.

Thus a relation between particles and dimensions is not unnatural in Physics.

9.1 Particles Mapped to Quantum Functionals

A particle has a quantum field. We have shown in earlier books that quantum fields may be factored into an inner product of a quantum functional and a wave in coordinate space-time.[49] Every particle has a set of internal symmetry and space-time quantum numbers. We can map the set of fundamental fermions to a set of quantum functionals. These functionals have transformation properties under internal symmetries. They are thus representations of the internal symmetry groups.

There is no distance in the set of functionals. As we showed in Blaha (2020) and earlier books factoring quantum fields into inner products of functionals and waves eliminates the instantaneity problem of Quantum Entanglement.

9.2 Coordinates and Functionals

The lack of distance in the set of functionals parallels the lack of distance between coordinates. For example there is no distance between the "x" coordinate and the "y" coordinate in a coordinate system. Functionals have features in common with coordinates.

---

[49] See Blaha (2019e) and (2020) as well as earlier books by the author.

Quantum functionals can also have commutation relations similar to coordinates: Defining a quantum functional conjugate momentum:

$$\pi_k = d/df_k \tag{9.1}$$

For quantum functional f_k we obtain the commutation relation

$$[f_k, \pi_j] = i\delta_{kj} \tag{9.2}$$

where k and j represent internal symmetry indices. Eq. 9.2 mirrors the form of quantum mechanical commutation relations adding to the similarity between coordinates (dimensions) and quantum functionals.

Biquaternion and bioctonion coordinates can be mapped to quantum functionals since they have the same role, and number, as the respective coordinates. We may set

$$x_m = R_{mn}f_n \tag{9.3}$$

where R is a transformation.

9.3 Particle-Dimension Duality

Thus in QUeST and MOST we find a particle-dimension (coordinate) duality. The point of view offered by this duality suggests that fermion particles are in some sense "interchangeable" with coordinates—with particle functionals as an intermediary.

9.4 Particle – Functional – Dimension Triality

The above discussion shows a more general analogy – a *triality* – between fermion particles, dimensions (coordinates) and quantum functionals.

9.5 Functionals Implement Quantum Entanglement

In 1935 Einstein, Podolsky, and Rosen[50] (EPR) raised questions about Quantum Entanglement.

9.5.1 Quantum Entanglement and Action-at-a-Distance

EPR considered the quantum entanglement of two systems and showed that instantaneous action-at-a-distance (spookiness) resulted. In this section we will show that our quantum functional formalism, which generalizes quantum theory, eliminates

---

[50] Einstein A, Podolsky B, and Rosen N, "Can Quantum-Mechanical Description of Physical Reality Be Considered Complete?", Phys. Rev. **47**, 777 (1935).

the problem of instantaneous action-at-a-distance.[51] We will show the solution provided by quantum functionals using the same example as EPR.

The key feature of quantum functionals is their ubiquitous presence at every point of space-time. In a multi-system quantum state all functionals are directly, instantaneously linked no matter what the separation of the constituent systems. When a reduction of the state of one system occurs due to a measurement, all other systems are instantly updated since the space-time separation of the individual systems is not relevant. The linkage of all quantum functionals is relevant. A reduction of one system immediately impacts the other related systems.

9.6 The EPR Two System State Example

EPR considered a state consisting of two systems that might become separated spatially. We can represent the state as

$$\Psi = \Sigma_n \psi_{1n}(x_1)\psi_{2n}(x_2) \tag{9.4}$$

We can represent a measurement (reduction of state) with a projection Π_{1a} of system "1" to a state ψ_{1a} with

$$\psi_{1a} = \delta_{ab} \, \Pi_a \, \psi_{1b} \tag{9.5}$$

Then

$$\Psi_{projected} = \Pi_{1a} \, \Sigma_n \psi_{1n}\psi_{2n} = \psi_{1a}(x_1)\psi_{2a}(x_2) \tag{9.6}$$

The effect of the measurement of system "1" is *instantaneous* of system "2" because the quantum functionals f_{1n} and f_{2n}, and the projections Π_{1n} and Π_{2n} of both systems are not separated by distance with the result

$$\psi_{1n}(x) = f_{1xn}(\Pi_{1xn}\Phi) = (f_{1xn}, \Pi_{1xn}\Phi) \tag{9.7}$$

$$\psi_{2n}(y) = f_{2ny}(\Pi_{2yn}\Phi) = (f_{2yn}, \Pi_{2yn}\Phi) \tag{9.8}$$

with $x = x_1$ and $y = x_2$. *The quantum functional and the projection select the wave and its coordinate parameterization. The coordinates in the wave are merely place holders.*

Therefore the relative distance between the coordinates x_1 and x_2 is not relevant for the change of state of system "2". The quantum functionals and projections give the instantaneity of the change in ψ_{2a} upon the measurement of system "1".

---

[51] This section presents the solution for quantum spookiness that we proposed in Blaha (2019g) and (2018e).

The EPR Spookiness is resolved by quantum functionals. There is no conflict with the Theory of Special Relativity.

Essence of Eternity II:
Quaternion Dimension – Fermion Duality in SuperStandard Theories

Stephen Blaha Ph. D.
Blaha Research

One-to-One Fermion-Dimension Duality Established
256 QUeST Fermions
Four Species of QUeST Fermions
Any Fermion is Transformable to Any Other Fermion Using Symmetry Groups
Particle Functionals Support Relativistic Instantaneous Entanglement
Particle Functionals, Monads, Observability
Observability and Absolute Reality

Pingree-Hill Publishing
MMXX

Copyright © 2020 by Stephen Blaha. All Rights Reserved.

This document is protected under copyright laws and international copyright conventions. No part of this book may be reproduced, stored in a retrieval system, or transmitted by any means in any form, electronic, mechanical, photocopying, recording, or as a rewritten passage(s), or otherwise, without the express prior written permission of Blaha Research. For additional information send an email to the author at sblaha777@yahoo.com or call 603-289-5435.

ISBN: 978-1-7345834-6-5

This document is provided "as is" without a warranty of any kind, either implied or expressed, including, but not limited to, implied warranties of fitness for a particular purpose, merchantability, or non-infringement.   This document may contain typographic errors, technical inaccuracies, and may not describe recent developments. This book is printed on acid free paper.

Rev. 00/00/01   July 4, 2020

INTRODUCTION

In previous books this author has derived the Unified SuperStandard Theory (UST) in our 3 + 1 dimension space-time from Complex General Relativity and Quantum Field Theory suitably extended. Recently we showed that 32 complex quaternion dimension QUeST gives the identical pattern of Internal Symmetries as UST. *UST is derivable from QUeST.*

This remarkable coincidence leads us to explore Unified SuperStandard Theories in greater detail.

In this book we examine particle-dimension duality and find that the one-to-one match extends down to the individual fermion and dimension level.

We then develop the monad – particle functional concept that had enabled us to eliminate the issues of instantaneous quantum entanglement with Special Relativity in earlier books.

We also defined a complex octonion space of 32 dimensions called MOST in previous books. It also has a close (fermion) particle-dimension duality.

These considerations led us to consider the implications of monads. We found it provides a basis for entanglement and observability. It leads to an absolute reality throughout the universe using particle functionals to give observability at every point in the universe. At each point of space-time gravitons, fermions and bosons provide obervability.

The Megaverse also has a close fermion-dimension duality. We find the Megaverse is also has total observability and absolute reality.

1. Dimensions of Unified SuperStandard Theories

Dimensions are usually thought to be static—existing only to be used to define the coordinates of physical theories. They are not thought to have a dynamic aspect. In this chapter we will define the dimensions of the Unified SuperStandard Theory (UST) in our space-time, and in an underlying complex quaternion space that we call QUeST. We will see that the internal symmetries of UST emerge directly from the set of dimensions of QUeST. Thus internal symmetry dimensions of UST and the Standard Model are no longer a subject of mystery and no longer unusual (as they are sometimes portrayed.)

The fundamental UST theory is described in detail in Blaha (2020c) (and earlier books. Its basis in 32 dimension complex quaternion space (QUeST) is described in Blaha (2020d) – also in detail. It is remarkable that QUeST provides an exact fundamental basis for UST.[52]

1.1 QUeST Dimensions

QUeST is defined in a 32 dimension complex quaternion space. There are a total of 256 individual dimensions in this space. In view of the number of dimensions the usual approach of defining coordinates is cumbersome. Consequently we followed a Mathematical Picture Language approach in Blaha (2020d). This approach was originally used by Pythagoras and His School around 500 BCE.[53] Pythagoras used • symbols, which he called *psiphi* symbols (meaning pebbles).

We will begin with an 8 dimension complex quaternion space which has 64 dimensions. Then we will consider the 32 dimension complex quaternion space which can be viewed as consisting of four layers of 8 dimension complex quaternion space.

---

[52] The basis of UST in QUeST was not known to the author until Fall, 2019 although the form of UST was known to the author many years earlier and recorded in several books.

[53] Kirk (1962) presents much of what is known of the Pythagoreans. This author developed the psiphi diagrams, originally, without being aware of the Pythagorean diagrammatic language.

We express the 8-dimension space as a diagram in Fig. 1.1. It consists of a pattern of psiphi. Then we will partition it into the dimensions of space-time and internal symmetry groups in the next section.

```
• • • •   • • • •
• • • •   • • • •
• • • •   • • • •
• • • •   • • • •
• • • •   • • • •
• • • •   • • • •
• • • •   • • • •
• • • •   • • • •
```

Figure 1.1. Psiphi diagram of the dimensions of 8 dimension complex quaternion space. Each row represents a complex quaternion with 8 dimensions.

Similarly the 32 dimension complex quaternion space is depicted as in Fig. 1.2.

```
• • • •   • • • •
• • • •   • • • •
• • • •   • • • •
• • • •   • • • •
• • • •   • • • •
• • • •   • • • •
• • • •   • • • •
   • • •
• • • •   • • • •
```

Figure 1.2. Psiphi diagram of the dimensions of 32 dimension complex quaternion space. Each row again represents a complex quaternion with 8 dimensions.

Fig. 1.1 is a psiphi diagram that represents 7 + 1 coordinates with 8 dimensions:

Time Biquaternion

$$t = (a + ib + jc + kd) + I(a' + ib' + jc' + kd') \qquad (1.1)$$

Spatial Biquaternions

$$x = (a_x + ib_x + jc_x + kd_x) + I(a'_x + ib_x' + jc_x' + kd_x')$$
$$y = (a_y + ib_y + jc_y + kd_y) + I(a'_y + ib_y' + jc_y' + kd_y')$$
$$z = (a_z + ib_z + jc_z + kd_z) + I(a'_z + ib_z' + jc_z' + kd_z')$$
$$x1 = (a_{x1} + ib_{x1} + jc_{x1} + kd_{x1}) + I(a'_{x1} + ib_{x1}' + jc_{x1}' + kd_{x1}')$$
$$y1 = (a_{y1} + ib_{y1} + jc_{y1} + kd_{y1}) + I(a'_{y1} + ib_{y1}' + jc_{y1}' + kd_{y1}')$$
$$z1 = (a_{z1} + ib_{z1} + jc_{z1} + kd_{z1}) + I(a'_{z1} + ib_{z1}' + jc_{z1}' + kd_{z1}')$$
$$w1 = (a_{w1} + ib_{w1} + jc_{w1} + kd_{w1}) + I(a'_{w1} + ib_{w1}' + jc_{w1}' + kd_{w1}')$$

where all coefficients: a, b, c, d, a', b', c', d', and a_i, b_i, c_i, d_i, a'_i, b'_i, c'_i, d'_i for i = x, y, z, w, x1, y1, z1, w1 are *real-valued* numbers, and where I is an additional fundamental quaternion unit that makes each quaternion "complex." Note that the real and imaginary part of each coordinate has the same fundamental quaternion units to permit complex rotations between them.

As we will see we need four iterations of the above set of complex quaternions for the space of QUeST so that it will yield UST upon restriction to real-valued coordinates. Thus QUeST requires a 32 dimension complex quaternion space:

$$t = (a + ib + jc + kd) + I(a' + ib' + jc' + kd') \qquad (1.2)$$
$$x = (a_x + ib_x + jc_x + kd_x) + I(a'_x + ib_x' + jc_x' + kd_x')$$
$$y = (a_y + ib_y + jc_y + kd_y) + I(a'_y + ib_y' + jc_y' + kd_y')$$
$$z = (a_z + ib_z + jc_z + kd_z) + I(a'_z + ib_z' + jc_z' + kd_z')$$
$$x1 = (a_{x1} + ib_{x1} + jc_{x1} + kd_{x1}) + I(a'_{x1} + ib_{x1}' + jc_{x1}' + kd_{x1}')$$
$$y1 = (a_{y1} + ib_{y1} + jc_{y1} + kd_{y1}) + I(a'_{y1} + ib_{y1}' + jc_{y1}' + kd_{y1}')$$
$$z1 = (a_{z1} + ib_{z1} + jc_{z1} + kd_{z1}) + I(a'_{z1} + ib_{z1}' + jc_{z1}' + kd_{z1}')$$
$$w1 = (a_{w1} + ib_{w1} + jc_{w1} + kd_{w1}) + I(a'_{w1} + ib_{w1}' + jc_{w1}' + kd_{w1}')$$

$$\cdots$$

$$w4 = (a_{w4} + ib_{w4} + jc_{w4} + kd_{w4}) + I(a'_{w4} + ib_{w4}' + jc_{w4}' + kd_{w4}')$$

1.2 Breakup of QUeST Space into Space-time and Internal Symmetry Groups

Returning now to psiphi diagrams we find that the 8-dimension complex quaternion space can be partitioned into blocks based on the dimension rules:

U(2) requires 4 dimensions
U(1)⊗SU(2) requires 4 dimensions
SU(3) requires 6 dimensions
U(4) requires 8 dimensions

where the dimensions have real-valued coordinates and are called *real dimensions*. Fig. 1.3 shows the partition of one layer QUeST (8 dimensions by eq. 1.1) giving eq. 1.3.

$$SU(2)\otimes U(1)\otimes SU(3)\otimes U(2)\otimes SU(2)\otimes U(1)\otimes SU(3)\otimes U(2) \qquad (1.3)$$

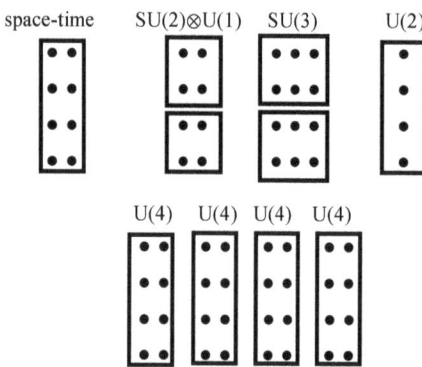

Figure 1.3. Psiphi diagram showing partitioning of 8 dimension complex quaternion space. The blocks of dimensions yield 4 complex dimension space-time, and the internal symmetries of normal matter and Dark matter SU(2)⊗U(1)⊗SU(3)⊗U(2)⊗SU(2)⊗U(1)⊗SU(3)⊗U(2) as they appear in UST. The lower U(4) groups are for the Generation group and the Layer group for normal matter and for Dark matter in one layer UST. The U(2) group transforms between normal and Dark matter.

1.3 Justification for a Four Layer QUeST

There is good reason for QUeST to have four layers embodied in 32 dimension complex quaternion space. If one considers the content of the layer displayed in Fig. 1.3 one sees a 4 dimension complex coordinates block for space-time. To create a 4 dimension complex quaternion coordinates space-time, one needs four layers of the form of Fig. 1.3. *The combination of the four 4-dimension complex coordinate parts is a complex quaternion dimensions space-time.* Thus the choice of four layer QUeST gives us a 4-dimension complex quaternion space-time AND enables QUeST to map directly to UST with its four layers if one limits the quaternion coordinates to the real-valued coordinates within them.[54] If one did not define a four layer QUeST then the role of the four U(4) Layer Groups would be in doubt since the Layer Groups in UST transform among each of the four generations of fermions. (See Fig. 1.6 for an illustration.) Four generations implies a need for four Layer groups, which the 32 complex quaternion QUeST contains. Note four U(2) groups that transform between normal and Dark sectors for each layer are required. We call these groups *Dark* groups.

We conclude four layer QUeST is needed to have a 4 dimension complex quaternion space-time.

1.4 One Layer QUeST Structure

As a preliminary to four layer 32 complex quaternion dimensions QUeST we display the structure implicit in Fig. 1.3 in Fig. 1.4.

---

[54] The Layer groups of UST enable mixing between the layers of fermions as shown in Fig. 2.1.

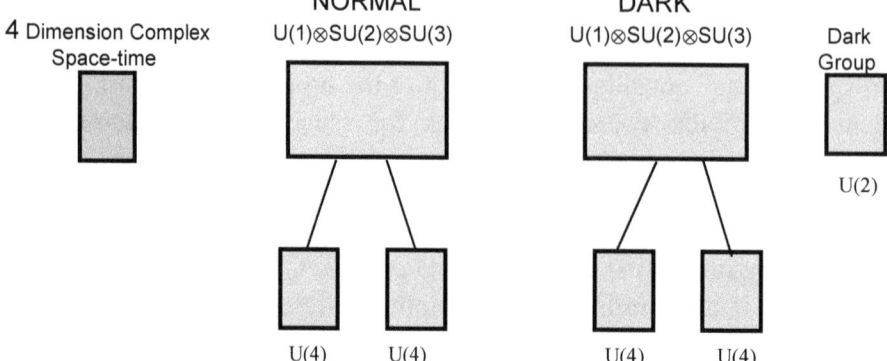

Figure 1.4. Schematic of the structure of the internal symmetry groups of eq. 1.3 plus 4 complex dimensions space-time. The two large blocks are each 5 dimension complex coordinate representations of SU(2)⊗U(1)⊗SU(3). The U(2) group supports transformations (rotations) between normal and Dark matter.

1.5 Four Layer QUeST

Fig. 1.5 shows four layer QUeST internal symmetry groups and 4 dimension complex quaternion space-time. Fig. 1.7 shows the 4 layer fundamental fermion spectrum. Fig. 3.1 shows the map between 32 dimension complex quaternion dimensions and fundamental fermions. The map is one-to-one for all four layers.

The 256 dimensions of the 32 dimension complex quaternion space equal the 256 fundamental fermions of QUeST and UST.

The internal symmetry group structure of Fig. 1.5 is

$$[SU(2) \otimes U(1) \otimes SU(3) \otimes SU(2) \otimes U(1) \otimes SU(3) \otimes U(4)^4 \otimes U(2)]^4 \qquad (1.4)$$

plus 4 dimension complex quaternion space-time.

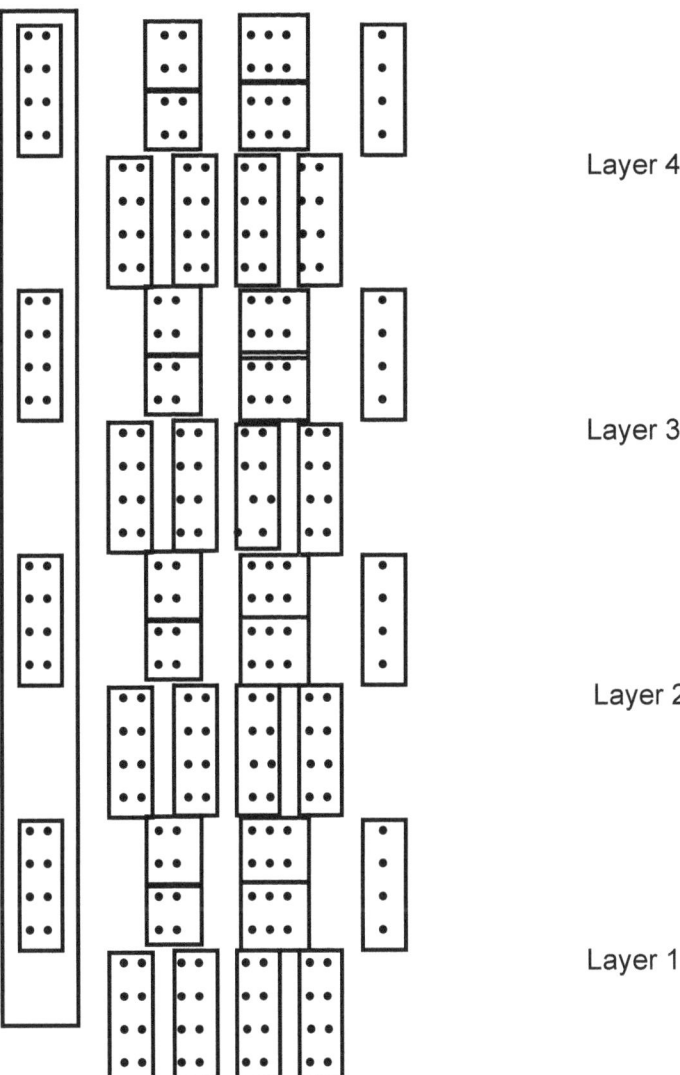

Layer 4

Layer 3

Layer 2

Layer 1

Figure 1.5.   Four layer QUeST internal symmetry groups and space-time diagram for 32 dimension complex quaternion space. Note the left composite blocks combine to specify a 4 dimension complex quaternion space-time.

1.6 Particles of UST and QUeST

In Blaha (2020c) and earlier books we found the set of internal symmetry groups of eq. 1.4 to which we added the Dark groups $U(2)^4$ based on a fundamental derivation of UST from QUeST in Blaha (2020a) through (2020c). QUeST provides a fundamental basis for UST and Standard Model internal symmetries removing the questions of strangeness often attributed to Standard Model symmetries.

The fundamental fermions of QUeST were found to be the same as in UST. Fig. 1.7 displays the four layers of fermions of UST and QUeST together with the roles of the $SU(2) \otimes U(1) \otimes SU(3)$ groups, the Generation groups (vertical within fermion generations), the Layer groups (vertical encompassing all four layers for each generation), the Dark groups (one-to-one, fermion by fermion between normal and Dark sectors), and the Complex Lorentz group. The role of these groups will be discussed in more detail in chapter 8.

There are 256 fundamental fermions counting quarks as triplets.

1.7 QUeST Vector Bosons

The overall *one layer* QUeST internal symmetry vector bosons are:

<u>"Normal" Gauge Groups</u>
$SU(3) \otimes SU(2) \otimes U(1)$
Generation Group U(4)
Layer Group U(4)

<u>Dark Gauge Groups</u>
$SU(3) \otimes SU(2) \otimes U(1)$
Generation Group U(4)
Layer Group U(4)

PLUS

A Dark U(2) group that rotates between the normal and Dark sectors

Figure 1.6. One layer QUeST vector bosons. The four layer QUeST quadruples the above list: with one distinct set for each layer.

The Fermion Periodic Table

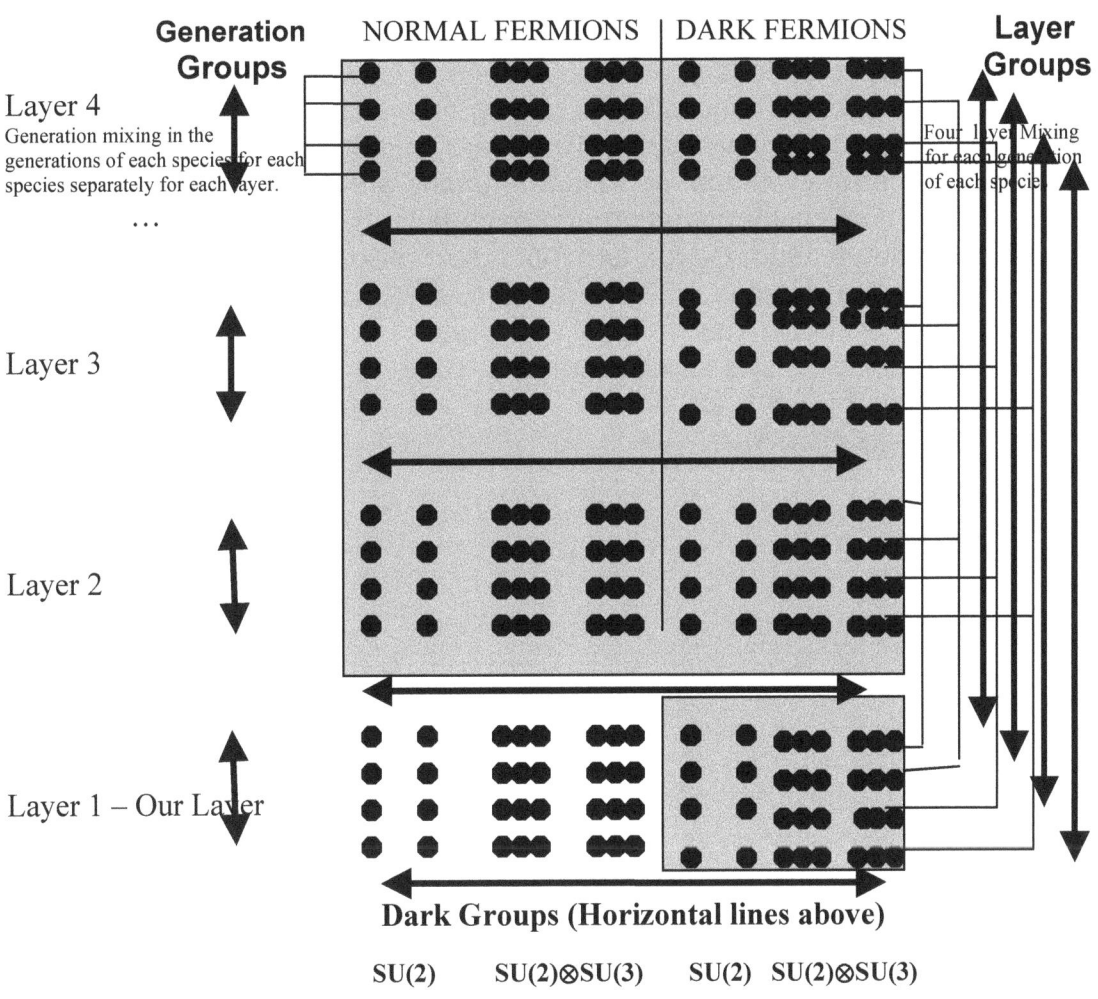

Figure 1.7. Fermion particle spectrum and partial example of pattern of mass mixing of the Generation, Layer, and Dark grroups. Unshaded parts are the

known fermions including an additional, as yet not found, 4th generation shown. The lines on the left side (only shown for one layer) display the Generation mixing within each layer's species. The Generation mixing applies within each layer using a separate Generation group for each layer. The lines on the right side show Layer group mixing with the mixing amongst all four layers for each of the four generations individually. There are four Layer groups. The Dark groups mixing between normal and Dark fermions are shown in the center as horizontal lines. For each generation and each layer SU(2) mixes between an e-type fermion and a neutrino-type fermion. It also mixes between an up-quark-type fermion and a down-quark-type fermion. SU(3) mixes among each up-quark triplet and down-quark triplet separately. Complex Lorentz group transformations map among all four fermions: Dirac ↔ tachyon ↔ up-quark ↔ down-quark. See Fig. 8.1 for details. There are 256 fundamental fermions counting quarks as triplets.

2. Megaverse 32 Complex Octonion Space (MOST)

The 32 complex octonion space of the Megaverse (or Multiverse) is factored into a 7 complex quaternion space-time and a set of internal symmetries.[55] The Magaverse can contain our universe as well as a host of other universes.

We find it convenient to split the Megaverse into four 8 complex octonion subspaces. These subspaces are duplicates of each other but contain different internal symmetry groups and different MOST fermion and vector boson spectrums.

Fig 2.1 symbolically depicts an 8-dimension complex octonion (bioctonion) space with a psiphi • for each real-valued dimension. *We treat bioctonion space as a higher dimensional space and do not use details of octonion algebra in our development.*

Figure 2.1. Eight-Dimensional (7 + 1) complex octonion subspace with coordinates represented by • 's. This subspace has 128 real dimensions.

The internal symmetry dimensions above number 112. Sixteen of the dimensions serve as the dimensions of an 8-dimension complex space-time. These

---

[55] Much of this chapter appears in Blaha (2020d).

dimensions serve as the fundamental representation dimensions[56] of each of the factors of

$$[SU(2)\otimes U(1)\otimes SU(3)\otimes SU(2)\otimes U(1)\otimes SU(3)]^2\otimes U(4)^9 \qquad (2.1)$$

counting a U(4) Dark group.

     The U(4) Generation and Layer groups are represented in Fig. 2.1. We depict the pattern of symmetry implied by Fig. 2.1 and eq. 2.1 in Fig. 2.2 and 2.3 below.

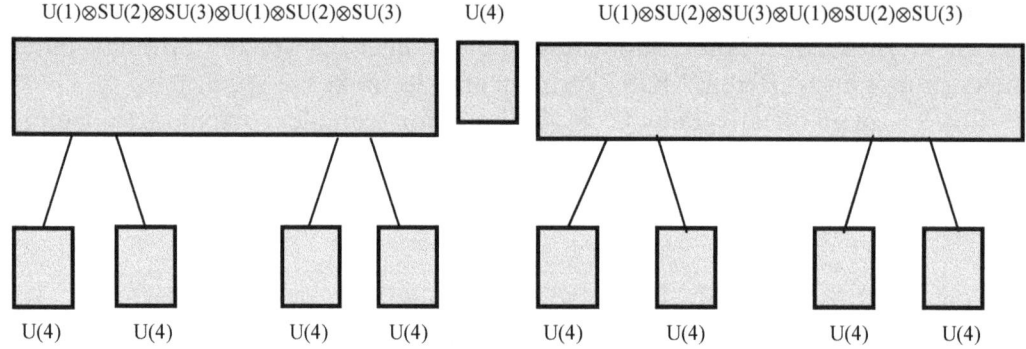

Figure 2.2. Schematic of the internal symmetry groups' dimensions of Fig. 2.1. The two "large" blocks are each sets of 20 real[57] dimensions furnishing representations of the indicated groups. The lower U(4) groups are the Generation and Layer number groups. The Dark U(4) group is shown. The total number of real dimensions is 112.

     Each U(1)⊗SU(2)⊗SU(3)⊗U(1)⊗SU(2)⊗SU(3) block in Fig. 2.3 has a 10 complex dimensions (20 real dimensions) representation. The blocks are subdivided in Fig. 2.3 into sets of 10 real dimensions supporting representations of U(1)⊗SU(2)⊗SU(3). We assign the first block to contain the representations of the known parts of the Standard Model. There are three Dark blocks. The internal symmetry groups of each part are listed in Fig. 2.3.

---

[56] See section 1.2.

[57] We also use the term complex dimensions to indicate pairs of real dimensions.

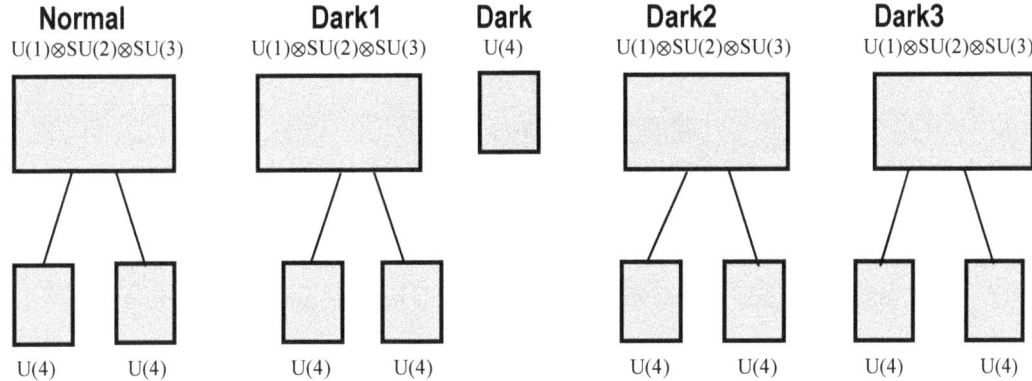

Figure 2.3. Schematic of the internal symmetry groups of eq. 2.1 including the Dark U(4) group. These are the internal symmetry groups of one layer MOST. The lower U(4) groups above are the Generation and Layer number groups. One pair of each number group is for each of the four U(1)⊗SU(2)⊗SU(3) factors above.

2.1 One Layer MOST

The above section specifies the *one layer* MOST. The symmetries of the three other layers are the same but their groups, and fermions, are individual to each layer. The groups of each layer can be flagged with a different index.

The overall one layer MOST internal symmetry is specified by Fig. 2.3, eq. 2.1, and an 8-dimension complex space-time. The internal symmetry groups of one Layer MOST are:

"Normal" Gauge Groups
SU(3)⊗SU(2)⊗U(1)
Generation Group U(4)
Layer Group U(4)
Dark1 Gauge Groups
SU(3)⊗SU(2)⊗U(1)
Generation Group U(4)
Layer Group U(4)
Dark2 Gauge Groups

SU(3)⊗SU(2)⊗U(1)
Generation Group U(4)
Layer Group U(4)

Dark3 Gauge Groups
SU(3)⊗SU(2)⊗U(1)
Generation Group U(4)
Layer Group U(4)

PLUS

A Dark U(4) group that rotates among the four normal and Dark sectors

Figure 2.4. One layer MOST vector bosons list from eq. 2.1. The four layer MOST quadruples the above list: with one distinct set for each layer. In one layer the total number of vector bosons of the above list is 192. Thus four layers yield a total count of 768 vector bosons in MOST (not counting the Species group which comes from General Relativity). We require each layer has a separate Dark U(4) rotation group.

2.2 Four Layer MOST

The four layer MOST is described by a 32 dimension complex octonion space. Thus it consists of four "copies" of the coordinates:

Figure 2.5. The 32 dimension MOST schematic. Four layer MOST has 512 real dimensions.

They yield four duplicates of the internal symmetry schematic in Fig. 2.3, and an 8 complex quaternion space-time consisting of 7 + 1 complex-valued quaternion coordinates (obtained by combining the four layers of 8 dimension complex space-time coordinates.)

The sum total of real dimensions is 512 as is the sum of the dimensions of the above parts constructed from the dimensions.

2.3 Justification for a Four Layer MOST

There is good reason for MOST to have four layers embodied in 32 dimension complex quaternion space. If one considers the content of a layer one sees a 8-dimension complex coordinates block for space-time. To create an 8 dimension complex *quaternion* coordinates space-time, one needs four layers. *The combination of four 8-dimension complex coordinates is an 8 dimension complex quaternion space-time.*

Thus the choice of four layer MOST gives us an 8-dimension complex quaternion space-time that can contain 4-dimension complex quaternion QUeST universes. *We conclude four layer MOST is needed to have an 8-dimension complex quaternion space-time.*

The thirty-two dimension complex octonion space contains an 8-dimension complex quaternion space-time and the four layers of Internal Symmetry groups shown in Fig. 7.5.

2.4 Fermion and Gauge Vector Boson Spectrums

The fermion and vector boson spectrums that emerge in MOST are those of an "enlarged" QUeST and Unified SuperStandard Theory. They are displayed below. MOST has an additional two Dark sectors beyond QUeST and the Unified SuperStandard Theory.

Vector Bosons

From Fig. 2.4 we find MOST has 192 vector bosons in one layer. Thus four layer MOST has a total count of 768 MOST vector bosons. There are two additional Dark vector boson sectors beyond QUeST and the Unified SuperStandard Theory.

Fermions

There are 512 fundamental fermions in MOST, which includes two additional Dark fermion sectors. Fig. 2.6 shows the MOST fermion spectrum.

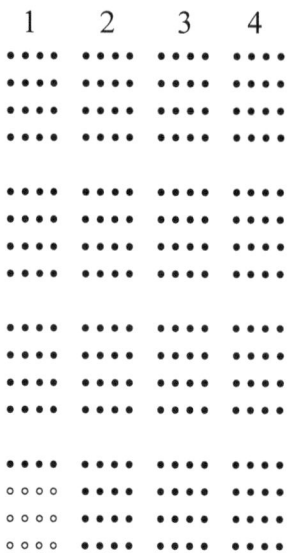

Figure 2.6. Schematic spectrum of the fermions of 4 layer MOST. Each fermion is represented by a •. Quark triplets are represented by a single •. Four sets of four species in four generations which are in turn in 4 layers. Open symbols ○ represent known fermions. There are 512 fundamental fermions taking account of quark triplets. Note the Layer groups determine the layers in UST. They require 4 layers of 8 complex octonions in Megaverse space leading to the 32 dimension complex octonion space.

3. Particle-Dimension Duality

3.1 Particle-Dimension Duality in our QUeST Universe

There is a remarkable correspondence between the dimensions of 32 dimension complex quaternion space and the fermion spectrum of UST and QUeST. Fig. 3.1 shows the map between the dimensions and fermions for one layer (8 complex quaternion dimensions and one layer of fermions in UST.) The other three layers of dimensions and fermions exhibit the same map. Consequently there is a one-to-one correspondence between the 256 dimensions and the 256 fundamental fermions in QUeST.

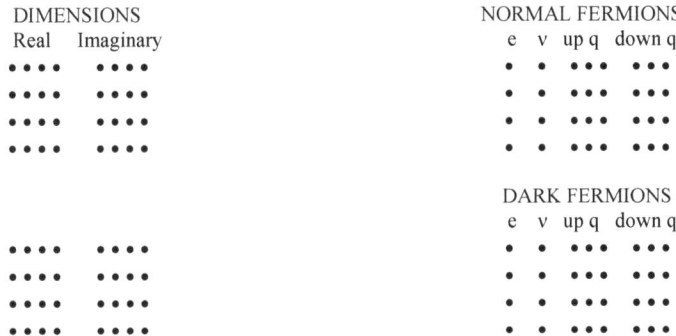

Figure 3.1. Schematic spectrum of the fermions of the one layer of normal and Dark fermions matching one layer of 8 rows of complex quaternion space dimensions on a one-to-one basis. The other three layers of space dimensions consisting of 24 complex quaternion dimensions are similar in having a one-to-one correspondence with three UST layers of fermions. UST has a four layer fermion spectrum. They total to 256 dimensions and 256 fundamental fermions.

The map is one-to-one and applies to individual particles and dimensions. Labeling the top row of dimensions of Fig. 3.1 as d_1, d_2, ..., d_8 we see the duality in detail for the top row of Fig. 3.1:

$$e \leftrightarrow d_1 \qquad\qquad (3.1)$$
$$\nu \leftrightarrow d_2$$
$$q_1 \leftrightarrow d_1 d_4$$
$$q_2 \leftrightarrow d_1 d_5$$
$$q_3 \leftrightarrow d_1 d_6$$
$$q_4 \leftrightarrow d_2 d_4$$
$$q_5 \leftrightarrow d_2 d_5$$
$$q_6 \leftrightarrow d_2 d_6$$

The quarks have both $SU(2) \otimes U(1)$ and $SU(3)$ symmetry resulting in the products of dimensions in the correspondence. (We assume free particles in these discussions.)

The one-to-one map applies to the rows of Fig. 3.1 for both normal and Dark sectors. Fig. 3.1 illustrates the following points:

1. A One-To-One Map: Dimensions – Fermions.

2. The 8 Complex Quaternion Dimensions in each of the top four rows correspond to normal $SU(2) \otimes U(1) \otimes SU(3)$ particles.

3. The 8 Complex Quaternion Dimensions in each of the lower four rows correspond to Dark $SU(2) \otimes U(1) \otimes SU(3)$ particles.

4. The 4 top Complex Quaternion rows corresponding to normal particles implies four fermion generations.

5. The 4 lower Complex Quaternion rows corresponding to Dark particles imply four Dark fermion generations.

6. 4 Layers of 8 Complex Quaternion rows imply 4 layers of fermions as seen in UST.

Thus our QUeST formalism accounts for the number of fermions per generation, the number of generations per layer (4) for normal and Dark fermions, the number of normal and Dark layers (4), the number of space-time dimensions (4), and the number of complex quaternion dimensions[58] (32).

3.2 Particle-Dimension Duality in our MOST Megaverse

There is also a one-to-one correspondence between the dimensions of 32 dimension complex octonion space and the fermion spectrum of MOST. Fig. 3.2 shows the map between the dimensions and fermions for one layer (8 complex octonion dimensions and one layer of fermions in MOST.) The other three layers of dimensions and fermions exhibit the same map. As a result there is a one-to-one correspondence between 512 dimensions and 512 fundamental fermions in MOST – the Unified SuperStandard Theory of the Megaverse.

3.3 Implications of Particle-Dimension Duality

Particle-dimension duality combined with the definition of particle fields in terms of functionals (See Blaha (2020c) and earlier books.), and a map of dimensions to coordinate dimensions of fundamental group representations, and thence to functionals enables us to define a triality of dimensions, functionals and particles. The triality may be symbolized by

$$d_i \leftrightarrow g_i \leftrightarrow f_i$$

where d_i is a dimension, g_i is the dimension taken to be a dimension of a group's fundamental representation, and f_i is a functional which is subject to the group's transformations.

We consider the implications of these considerations in succeeding chapters.

---

[58] Eq. 1.2.

Top Four Rows of One Layer of 8 Complex Octonions

DIMENSIONS   NORMAL FERMIONS
  Real Part   e v up q down q
• • • • • • • •   • • • • • • • •
• • • • • • • •   • • • • • • • •
• • • • • • • •   • • • • • • • •
• • • • • • • •   • • • • • • • •

                                                DARK1 FERMIONS
  Imaginary Part   e v up q down q
• • • • • • • •   • • • • • • • •
• • • • • • • •   • • • • • • • •
• • • • • • • •   • • • • • • • •
• • • • • • • •   • • • • • • • •

Lower Four Rows of One Layer of 8 Complex Octonions

DIMENSIONS   DARK2 FERMIONS
  Real Part   e v up q down q
• • • • • • • •   • • • • • • • •
• • • • • • • •   • • • • • • • •
• • • • • • • •   • • • • • • • •
• • • • • • • •   • • • • • • • •

                                                DARK3 FERMIONS
  Imaginary Part   e v up q down q
• • • • • • • •   • • • • • • • •
• • • • • • • •   • • • • • • • •
• • • • • • • •   • • • • • • • •
• • • • • • • •   • • • • • • • •

Figure 3.2. Schematic spectrum of the fermions of the one layer of normal and Dark fermions matching one layer of 8 rows of complex octonion space (See Fig. 2.1) on a one-to-one basis. The other three layers of space dimensions consisting of 24 complex octonion dimensions are similar in having a one-to-one correspondence with three UST layers of fermions. UST has a four layer fermion spectrum. They total to 512 dimensions and 512 fundamental fermions.

4. Status of Elementary Particle Theory

Elementary particle theory has gone through a number of phases in the past forty years. Forty years ago model quantum field theories largely based on quantum field theory and various forms of internal symmetries. The motivation for these theories was the success of the Standard Model. It was thought more was needed.

The resulting theories incorporated the Standard Model within generalizations of Standard Model internal symmetries. The primary issue was the justification for proposed internal symmetries.

There was also a deeper approach based on SuperString theory. SuperString theories encountered numerous issues. A primary issue was the selection of a SuperString theory that led inexorably to a physically realistic theory along the lines of the Standard Model. The search for the "true" SuperString theory, if it exists, has occupied the efforts of a sizeable number of theorists for forty years.

In view of the quagmire of theoretic efforts for a Theory of Elementary Particles this author developed a theoretic approach based on a set of fundamental postulates in the manner of Euclid. With six postulates the author was able to develop the known parts in the form of the Standard Model. Part of this development was the introduction of a modified quantum field theory formalism using Two Tier field theory (to eliminate divergences in perturbation theory) and PseudoQuantum field theory which supports higher derivative lagrangians in a canonical manner and enables reasonable physics in curved space-times. It also appeared reasonable to base the theory on Complex Special Relativity. Complex General Relativity became necessary in order to create a unified theory for curved space-time. The result was the Unified SuperStandard Theory (UST).

4.1 Unified SuperStandard Theory (UST)

UST had a fundamental fermion spectrum that contained the known fermion spectrum, and added a fourth generation, added four layers of fermions, and added a Dark matter spectrum of fermions that mirrored the normal fermion spectrum. The vector boson interactions contained the ElectroWeak interactions, U(4) interactions that caused interactions between fermions in each layer and similarly for Dark matter. The Generation group interactions yielded $U(4)^8$. The theory also had interactions between the layers of fermions.

UST also contained the Higgs boson sector and gravitational interactions according to a modified General Relativity. General Relativity and the Strong interactions were modified to have higher order derivative field equations. Consequently the gravitational force was modified in a manner similar to MoND. Standard Model interactions were modified to have a linear potential.

The theory is described in Blaha (2018f) and (2020c).

4.2 Quaternion Unified SuperStandard Theory (QUeST)

QUeST was defined in 32 dimension complex quaternion space. The study of QUeST[59] showed that it had the internal symmetries of UST plus an additional Dark U(2) interaction that transformed between normal particles and their Dark equivalents for all four fermion layers giving an additional $U(2)^4$ symmetry.

The space-time generated from QUeST a 4-dimension complex quaternion space that became real 4-dimension space-time in UST when quaternions were restricted to real coordinates.

Thus we saw that UST is properly viewed as derivative from QUeST. This unanticipated result solidified our feeling that QUeST and UST are the true theories of elementary particles.

4.3 Potential QUeST Perturbation Theory Divergences

QUeST is defined in a 4-dimension complex quaternion space-time. Just as 4-dimension complex space-time is restricted to real-valued 4-dimension space-time, QUeST is restricted to the 4-dimension real-valued space-time of UST.

The question then arises of high energy infinities in quaternion space-time since it has 32 real-valued dimensions. Integrations in momentum space in this space-time have the general form

$$\int d^{32}k$$

---

[59] See Blaha (2020a) through (2020d)/

Since particle propagators tend to be inverse quadratic (or inverse cubic) it appears that perturbation theory integrations will be highly ultraviolet divergent.

However if one uses Two-Tier coordinates as we do in UST an exponentially convergent factor in all dimensions appears eliminating ultra-violet divergences. In Two-Tier quaternion theory coordinates are replaced with Two-Tier coordinates

$$X^\mu - x^\mu + iY^\mu / M^2$$

for i = 1, 2, …, 32 where M is a very large mass of the order of the Planck mass presumably. Thus the Two-Tier version of QUeST has convergent perturbation theory integrals.

4.4 Consequences

The smooth transition from QUeST to UST increases the likelihood that QUeST is the true theory of elementary particles and gravitation. It encompasses all we know experimentally. And it has the ability to grow as more is learned about Higgs symmetry breaking.

If QUeST is the correct theory then elementary particle physics become similar to Chemistry. However there is still much to learn about detailed features and much room for the expansion of the understanding of the basis of the theory.

4.5 MOST Extension to the Megaverse/Multiverse

A remarkable aspect of the theoretical approach presented in Blaha (2020c), and here, is its ability to be extended to a Megaverse/Multiverse of universes. MOST showed that a 32 dimension complex octonion space leads to a 8 dimension complex quaternion space-time and an enlarged set of internal symmetries. Thus our universe, and unlimited numbers of additional universes, can "fit" into the MOST Megaverse. The internal symmetries of QUeST are a subset of the internal symmetries of MOST — a salutary feature.. Thus we have a fairly complete total picture of the universe and of the Megaverse.

4.6 Future Directions

Perhaps the most important issue facing particle theory is to obtain an understanding of the rationale for basing QUeST on complex quaternions. We know that Streater (2000) was able to justify extending space-time to complex-valued coordinates. What justification is there for complex quaternion coordinates? The number of these dimensions is 32. A partial justification for 32 dimensions is the need to obtain a 4-dimension complex quaternion space-time from which our real-valued space-time emerges.

Other tasks that remain are:

1. Determine the symmetry breaking for the internal symmetries including the symmetry breakdown(s) of 32 dimension complex quaternion space to internal symmetries and space-time.

2. Find the missing fermions and their interactions.

3. Determine the full set of symmetry breakdown parameters.

4. Find the deeper meaning of Higgs symmetry breaking from vacuum dynamics.

5. Develop a Chemistry of elementary particles.

5. Particle Functionals

5.1 Quantum Entanglement and Action-at-a-Distance

In 1935 Einstein, Podolsky, and Rosen[60] (EPR) raised questions about Quantum Entanglement. EPR considered the quantum entanglement of two systems and showed that instantaneous action-at-a-distance (spookiness) apparently resulted.

In this chapter we show how instantaneity is resolved by "factoring" wave functions into the composition of a functional and a Fourier wave function expansion. in In section 5.3 we show the instantaneity of the EPR two system state example is eliminated using functional factoring of wave functions.

Then we establish a functional formulation of the internal symmetry dimensions of 32 complex quaternion space. The functional formulation then generates fundamental representations of internal symmetry groups.

5.2 Functional Factorization of Quantum Fields

Many years ago Dirac factored the Klein-Gordan equation and obtained the Dirac equation for spin ½ fermions.

In the following sections we show that there is good reason to factor quantum mechanical wave functions and second quantized fields into an inner product of a particle functional and a corresponding Fourier coordinate expansion. With this factorization, and the assumption that the space of all particle functionals, as well as the space of all Fourier coordinate expansions, are both point spaces[61] with the consequence that there is no distance measure in either space, we find a change in one of a pair of space-like separated parts of an initial state causes an instantaneous transformation of the other part (eliminating Einstein's spookiness).

---

[60] Einstein A, Podolsky B, and Rosen N, "Can Quantum-Mechanical Description of Physical Reality Be Considered Complete?", Phys. Rev. **47**, 777 (1935).

[61] The points can be viewed as spaces with no distance measure that are factors in a tensor product with space-time.

5.3 Motivation for Quantum Field Factorization: Instantaneous Effects in Quantum Phenomena

Seemingly instantaneous quantum phenomena are apparent in many cases. For example:

1. Two particles placed in a definite spin state may separate to a space-like distance. If the z component of spin is flipped in one of the particles, the other particle instantaneously flips its spin in such a way as to conserve spin. This type of phenomena has been described as 'spooky' since it violates the law that no effect can travel at a rate faster than the speed of light.

2. Transitions between atomic levels take place instantaneously—in a zero time interval.

Quantum field factorization enables instantaneous effects to happen without violating Relativity.

5.4 General Form of Factorization

Normally fermion and boson quantum fields are described by a wave function of the form

$$\chi(\mathbf{x}, t) \tag{5.1}$$

We can formally factorize quantum fields as an inner product of a functional f_k and a space-time Fourier expansion denoted $(k, \mathbf{x}, t)$ (neglecting internal quantum numbers temporarily) where k is the momentum.

$$\chi(\mathbf{x}, t) = (f_k, (k, \mathbf{x}, t)) \tag{5.2}$$

For a free *two* particle wave function (non-interacting) the wave function may be written as a product of inner products:

$$\chi(\mathbf{x}, t) = (f_{1k}, (k, \mathbf{x}, t)_1) (f_{2q}, (q, \mathbf{x}, t)_2) \tag{5.3}$$

where k and q are momenta.

5.5 Factorization of Fermion Quantum Fields

We now consider the example of a free fermion field to illustrate the general concept. We began by defining a coordinate space Dirac Fourier quantum expansion as

$$(s, x, t) = N(p)[b(p, s)u(p, s)e^{-ip \cdot x} + d^\dagger(p, s)v(p, s)e^{+ip \cdot x}] \qquad (5.4)$$

where $N(p)$ is a normalization factor, u and v are functions of spin and momentum, and b and $d^\dagger$ are creation/annihilation operators. We defined a Dirac quantum wave function with the inner product of a functional and a coordinate space Fourier quantum expansion:

$$\psi(x) = (f, (s, x, t)) = \sum_{\pm s} \int d^3p N(p)[b(p, s)u(p, s)e^{-ip \cdot x} + d^\dagger(p, s)v(p, s)e^{+ip \cdot x}] \qquad (5.5)$$

where we use a functional inner product formalism in the manner of Riesz (1955)[62] and others.

5.6 The EPR Two System State Example

EPR considered a state consisting of two systems that might become separated spatially. We can represent the state as

$$\Psi = \sum_n \psi_{1n}(x_1)\psi_{2n}(x_2) \qquad (5.6)$$

We can represent a measurement (reduction of state) with a projection Π_{1a} of system "1" to a state ψ_{1a} with

$$\psi_{1a} = \delta_{ab} \Pi_a \psi_{1b} \qquad (5.7)$$

Then

$$\Psi_{projected} = \Pi_{1a} \sum_n \psi_{1n}\psi_{2n} = \psi_{1a}(x_1)\psi_{2a}(x_2) \qquad (5.8)$$

---

[62] For example see pp. 61-2 of Riesz (1955) where linear functionals and their inner products are defined.

The effect of the measurement of system "1" is *instantaneous* of system "2" because the quantum functionals f_{1n} and f_{2n}, and the projections Π_{1n} and Π_{2n} of both systems are not separated by distance with the result

$$\psi_{1n}(x) = f_{1xn}(\Pi_{1xn}\Phi) = (f_{1xn}, \Pi_{1xn}\Phi) \qquad (5.9)$$

$$\psi_{2n}(y) = f_{2ny}(\Pi_{2yn}\Phi) = (f_{2yn}, \Pi_{2yn}\Phi) \qquad (5.10)$$

with $x = x_1$ and $y = x_2$. *The quantum functional and the projection select the wave and its coordinate parameterization. The coordinates in the wave are merely place holders.*

Therefore the relative distance between the coordinates x_1 and x_2 is not relevant for the change of state of system "2". The quantum functionals and projections give the instantaneity of the change in ψ_{2a} upon the measurement of system "1".

The EPR Spookiness is resolved by quantum functionals. There is no conflict with the Theory of Special Relativity.

5.7 Factorization Details

The rationale for factorization lies in the nature of the functionals and coordinate Fourier expansions that we use. For, we choose to create a space of particle functionals for fermions, bosons, and other particle states that consists of a single point with no distance measure (or alternately put, zero distance between all functionals.) We also choose to create a 'point' space of all coordinate Fourier expansions for bosons and fermions, whose elements have all coordinate values, x.

For the moment we wish to note that the space of functionals includes functionals for all fundamental particles, and all matter/energy composites, in the universe (and the Megaverse). We can describe transitions (interactions) in which functionals are "transformed" into other functionals. So the space of functionals has a dynamic aspect. Another important aspect of functional space is its universality—*all functionals of the Megaverse are present creating a type of link between all parts of the Cosmos.*

The space of coordinate Fourier expansions consists of all possible expansions for particles in the coordinates of each respective universe and of the Megaverse. This space also has no distance measure.

The factorization that we propose enables instantaneous communication of a transition between two space-like separated parts of a state. A change in one part immediately causes a corresponding change in the other part because the changes take place in the functionals which are located at the same point in functional space.

In a certain sense we have divorced quantum phenomena from coordinate space by quantum field factorization.

6. Functionals in QUeST's 32 Dimension Complex Quaternion Space

The dimensions of QUeST space can be initially treated simply as independent dimensions. Dimensions can be mapped to particles and thence to their particle functionals. These functionals can then be used to furnish fundamental representations of internal symmetry groups. They can also be used to define quantum fields for elementary particles as in eq. 3.1. Since some fundamental fermions, namely quarks, have transformation properties of both SU(3) and SU(2)⊗U(1) the quark functionals are equivalent to composites of dimensions. Thus eq. 3.1 becomes

Particle Functional		Dimension Composite	
e	$\leftrightarrow$	d_1	(6.1)
ν	$\leftrightarrow$	d_2	
q_1	$\leftrightarrow$	$d_1 d_4$	
q_2	$\leftrightarrow$	$d_1 d_5$	
q_3	$\leftrightarrow$	$d_1 d_6$	
q_4	$\leftrightarrow$	$d_2 d_4$	
q_5	$\leftrightarrow$	$d_2 d_5$	
q_6	$\leftrightarrow$	$d_2 d_6$	

for the SU(3)⊗SU(2)⊗U(1) set of fermions in each generation and each layer of normal and Dark fermions.

6.1 Form of a Fundamental Fermion Particle Functional

The general form of a fermion functional is

$$F_{fermion} = f_{internal} f_{spin} \qquad (6.2)$$

where $f_{internal}$ is labeled with internal symmetry quantum numbers and f_{spin} is labeled by the spin state. They are factored to avoid SU6)-like problems found in the 1960s.

We will call functionals of the form of $F_{fermion}$ *fermion particle functionals*. We will call functionals of the form of $f_{internal}$ *internal symmetry functionals* since they embody internal symmetries. We will call functionals of the form of f_{spin} *spin functionals* since they embody spin. In chapter 7 we will consider boson functionals.

6.2 Particle Functional Byte Numbering

The 256 fundamental fermions may be numbered from 1 through 256. The ASCII character tables have an 8-bit (byte) numbering. We may then represent a particle functional as

$$f_{byte} = f_{internal} \qquad (6.3)$$

using bytes with values from 1 through 256.

Consequently fermions can be characterized as "letters" in an alphabet, and aggregates of fermions can be viewed as words. As a result one may think of the universe as a great Word—a concept which has been put forward many times.[63]

---

[63] A study of this possibility appears in Blaha (1998).

7. Four-Dimension Complex Quaternion Space-Time

QUeST is a theoretical foundation for the author's Unified SuperStandard Theory (UST). It is based on a 32 complex quaternion space with a total of 256 dimensions. Within this space there are a 4 dimension complex quaternion space-time with 32 dimensions, and a 224 dimension Internal Symmetry sector containing groups of UST including the U(2) Dark group..

The 4 dimension complex quaternion space-time is broken to a 4 complex dimension space-time, which becomes our space-time upon restriction to real dimensions (coordinates).

---

4 Dimension Complex Quaternion → 4 Dimension Complex → 4 Dimension Real

---

Figure 7.1. Progression of Space-times from QUeST to UST.

8. Complex Lorentz Group Boosts and Four Fermion Species

8.1 Complex Lorentz Boosts to Obtain the Four Species of Fermions

In UST[64] we showed that there are four Complex Lorentz group boosts that transform a spinor of a fermion at rest to four possible states: a Dirac electron-like fermion, a tachyon neutrino-like fermion, an up-quark-like fermion with real energy and complex 3-momentum, and a down-quark-like tachyon fermion with real energy and complex 3-momentum. We identified these fermions with the known leptons and quarks. We called each type of fermion a *species*. We showed that there was evidence to support faster-than-light neutrinos, and no evidence to prove down-type quarks were not tachyons. (In fact known fits to deep inelastic electron-nucleon data involve cutoffs that might be due to tachyon down quarks.)

In this chapter we examine the 4 dimension complex quaternion space-time of QUeST[65] and the complex space-time of UST (before restriction to real coordinates.) Each yields the four species of fermions.

The UST derivation of the four fermion species appears in Appendix A.

8.2 The Four QUeST Fermion Species

QUeST is defined in a 4-dimension complex quaternion space. We have extracted four complex-valued (eight real-valued) coordinates to form a space-time in chapter 7. Complex coordinates support a Complex Lorentz group just as the United SuperStandard theory does to some degree.

Appendix A describes the origin and features of the four species of fundamental fermions in UST in detail: "charged" lepton species, neutral lepton species, up-type quark species, and down-type quark species.

The features of the QUeST fermion species are the same as in UST. The key relation in complex quaternion space is

---

[64] Blaha (2020c) and earlier books such as Blaha (2007b).
[65] We also discuss the four species of Megaverse MOST.

$$e^x = e^a (\cos (\|\mathbf{v}\|) + \mathbf{v}/\|\mathbf{v}\| \sin(\|\mathbf{v}\|))s \qquad (8.1)$$

It is analogous to the similar complex coordinates identity used in Appendix A in eqs. 3.2, 3.3, 3.11, 3.12 and so on With it we can determine the boosts in 4 dimension complex quaternion space-time

8.3 QUeST Fermion Species

The fermion species in QUeST number four despite its four-dimension complex quaternion coordinates. It supports a 3+1 dimension Complex Quaternion Lorentz group. The cause is the central role of the speed of light c in QUeST (and in UST and MOST). We set c = 1. A massless particle travels at the speed of light in QUeST in the 24 spatial dimensions of QUeST, and satisfies $p^{02} - \|\mathbf{p}\|^2 = 0$ where $\|\mathbf{p}\|$ is given by eq. 8.2 below. The speed of light separates the boosts of the Complex Quaternion Lorentz group into four types.

The four types of boosts in this 3+1 dimension space-time that boost a particle rest state to a state of motion with a real-valued energy[66] and a real-valued or complex quaternion spatial momentum are:[67]

1. A boost from rest to a frame with real-valued energy and momentum with $p^{02} - \|\mathbf{p}\|^2 > 0$. A "normal charged lepton-like" fermion.

2. A boost from rest to a frame with real-valued energy and momentum with $p^{02} - \|\mathbf{p}\|^2 < 0$. A "tachyonic neutral lepton-like" fermion.

3. A boost from rest to a frame with real-valued energy and a complex quaternion valued spatial momentum with $p^{02} - \|\mathbf{p}\|^2 > 0$. An "up-type quark-like" fermion.

---

[66] The energy must be real-valued to have a stable (in the absence of interactions) fundamental particle. In 3 + 1 real-valued space-time complex energy implies a "resonance" that decays.

[67] QUeST and MOST space-times must have a speed of light which we will denote as c. The speed of light distinguishes between "normal" and tachyon particles.

4. A boost from rest to a frame with real-valued energy and a complex quaternion valued spatial momentum with $p^{02} - \|\mathbf{p}\|^2 < 0$. A "tachyonic down-type quark-like" fermion.

where p^0 is the energy and $\|\mathbf{p}\|$ is the norm of the complex quaternion valued spatial 3-vector $\mathbf{p}$ with the form in 3+1-dimension complex quaternion space-time:

$$\mathbf{p} = \mathbf{p_r} + i\mathbf{p_i} + j\mathbf{p_3} + k\mathbf{p_4} + q\mathbf{p_5} + r\mathbf{p_6} + s\mathbf{p_7} + t\mathbf{p_8} \qquad (8.2)$$
$$\|\mathbf{p}\| = \mathrm{sqrt}(\mathbf{p_r}{\cdot}\mathbf{p_r} + \mathbf{p_i}{\cdot}\mathbf{p_i} + \mathbf{p_3}{\cdot}\mathbf{p_3} + \ldots + \mathbf{p_8}{\cdot}\mathbf{p_8})$$

where i, j, k, q, r, s, and t are fundamental quaternion units, and where each momentum component $\mathbf{p_k}$ is a 3-vector. The terms like $\mathbf{p_r}{\cdot}\mathbf{p_r}$ are 3-vector inner products.

In UST the momentum has the general form

$$\mathbf{p} = \mathbf{p_r} + i\mathbf{p_i} \qquad (8.3)$$

See Appendix A for details.

Normal and tachyonic particles are distinguished in 4 dimension complex quaternion space-time by the sign of $p^{02} - \|\mathbf{p}\|^2$ in a manner similar to 4-dimension complex space-time. QUeST fermions occur in four species just like UST fermions.

8.4 MOST Fermion Species

In the Megaverse, the fermion species in MOST number four despite the 8-dimension complex quaternion nature of the space-time extracted from 32 dimension complex octonion space. The eight complex quaternion space-time coordinates support a 7+1 dimension Complex Quaternion Lorentz group.

There are four types of boosts in this 7+1 complex quaternion valued space-time that boost a particle rest state to a state of motion with a real energy, and a real-valued or complex quaternion valued momentum (spatial coordinates):

1. A boost from rest to a frame with real-valued energy and momentum with $p^{02} - \|\mathbf{p}\|^2 > 0$. A "normal charged lepton-like" fermion.

2. A boost from rest to a frame with real-valued energy and momentum with $p^{02} - \|\mathbf{p}\|^2 < 0$. A "tachyonic neutral lepton-like" fermion.

3. A boost from rest to a frame with real-valued energy and a complex quaternion valued spatial 7-momentum with $p^{02} - \|\mathbf{p}\|^2 > 0$. An "up-type quark-like" fermion.

4. A boost from rest to a frame with real-valued energy and a complex quaternion valued spatial 7-momentum with $p^{02} - \|\mathbf{p}\|^2 < 0$. A "tachyonic down-type quark-like" fermion.

where p^0 is the energy, and where each momentum component $\mathbf{p}_k$ is a spatial 7-vector:

$$\mathbf{p} = \mathbf{p}_r + i\mathbf{p}_i + j\mathbf{p}_3 + k\mathbf{p}_4 + q\mathbf{p}_5 + r\mathbf{p}_6 + s\mathbf{p}_7 + t\mathbf{p}_8 \qquad (8.4)$$
$$\|\mathbf{p}\| = \mathrm{sqrt}(\mathbf{p}_r \cdot \mathbf{p}_r + \mathbf{p}_i \cdot \mathbf{p}_i + \mathbf{p}_3 \cdot \mathbf{p}_3 + \ldots + \mathbf{p}_8 \cdot \mathbf{p}_8)$$

where i, j, k, q, r, s, and t are fundamental quaternion units, and where $\|\mathbf{p}\|$ is the norm of $\mathbf{p}$. The terms like $\mathbf{p}_r \cdot \mathbf{p}_r$ are 7-vector inner products.

8.5 Transitions Between Fermions in Different Species

Fermions appear in species, in generations, and in layers. (See Fig. 1.7) Generation groups'[68] transformations can transform a given fermion to a corresponding fermion in another generation within the same layer. Layer groups'[69] transformations can transform a fermion to the corresponding fermion in a different layer (while retaining the species and generation identity.) Dark groups'[70] transformations can transform between corresponding fermions in the normal and Dark fermion sectors.[71]

Transformations between fermions in different species in the same generation and layer require the use of SU(2)⊗U(1) and SU(3) as well as Complex Lorentz group transformations. See Fig. 8.1.

The role of each group in fermion transformations between species is:

---

[68] Note there are eight Generationr groups: one for each layer for both the normal and Dark fermion sectors.

[69] Note there are eight Layer groups: one for each of the four generations for both the normal and Dark fermion sectors.

[70] Note there are four Dark groups: one for each of the four layers.

[71] We treat the set of fermions as free in these discussions.

1. SU(2) – transforms between e and ν, and between up-quark and down-quark based on the "charged" SU(2) generators.
2. SU(3) – transforms among the three up-quarks, and transforms among the three down-quarks.
3. Complex Lorentz Group – transforms among each of the four species using transformations in section 8.3 (and Appendix A). Complex Lorentz group transformations map among all four fermion species: Dirac ↔ tachyon ↔ up-quark ↔ down-quark.

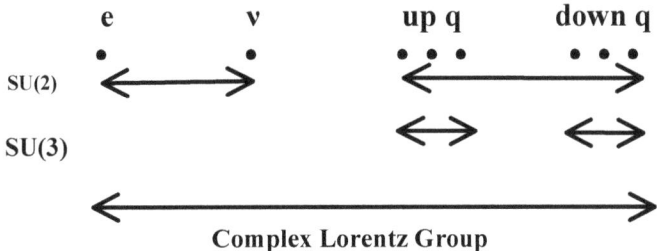

Figure 8.1. Transformations between fermions in differing species (and colors for quarks) for each generation in each layer.

The result of the set of all UST transformations (and similarly for QUeST transformations, which are analogous) is:

Any fermion can be transformed to any other fermion.

There are no isolated fermions. In the 256 fermion UST and QUeST spectrums.

As a result all fermions can be generated from any one fermion by applying the above transformations. Bearing in mind dimension-fermion duality we can view the dimension dual of the one fermion as the "source" of fermions and corresponding dimensions with the other 255 particles and dimensions being generated by transformations parallel to the above $SU(2) \otimes U(1) \otimes SU(3)$ and Complex Lorentz group transformations.

9. Particle Functionals Features

9.1 The Logic Core of Fundamental Fermions and Bosons

In previous books we opened the possibility that fermions (and bosons) might have a core that embodies logic in the form of spin as well as bare masses in the case of fermions. We defined functionals of various spins: 0, ½, 1, and 2. We saw that the core of spin ½ fermion functionals (that we called *qubes* in analogy with *qubits*) have a bare mass that we denoted m_0.

Bosons have cores as well that are boson functionals. Analogously we called a boson core a *quba*[72]. Boson functionals are massless. Bosons acquire masses through interactions.

The rationales for logic cores for particles is discussed in detail in chapters 3 and 8 of Blaha (2018e) and (2020c). We showed that the formalism based on a space of all particle functionals can lead to an explanation of the 'spooky' action at a distance of Quantum Entanglement that has been the subject of much discussion.

9.1.1 The Logic Building Block of Fermions – Qube Cores

If we consider all possible 'things' that might constitute a fundamental building block for a fundamental fermion theory they are all, at best, *ad hoc* and raise questions of their necessity and whether they are composed of yet a more fundamental substructure.

There is only one choice of building block that avoids these issues – a logic unit or qubit. A qubit is a fundamental entity that is a complex form of computer bit. A bit (and thus a qubit) is known to have an energy, or equivalently a mass, and has no constituents of a more primitive form.[73] We call a unit of logic that forms the core of a

---

[72] We use 'quba' simply because of its similarity to 'qube'. The leading 'b' signifies its bosonic use. We pronounce 'quba' as 'bub' with a silent 'e.' The word 'quba', itself, is the name of a Bantu language spoken by the Bubi people of Bioko Island in Equatorial Guinea.

[73] A qube is a physical manifestation of a logical value. The relation of a qube to a logical value is analogous to the relation of a penciled point placed on paper to the concept of a point as a primitive in geometry.

particle a *qube*.[74] It exists as the core of a particle. But, in itself, it has no *independent* material existence or space-time coordinates. A qube is a functional that acquires features such as coordinates, through functional inner products to become an elementary particle. We define a qube as a fermion field theory functional. (See chapters 3 and 8 of Blaha (2018e) or (2020c).)

9.1.2 Mass of a Qube

Recent experiments have shown that a logical value of a qubit has an energy associated with it. One bit of information has about 3×10^{-21} joules of energy[75] or a rest mass, m_0, or about 0.02 eV using $m_0 = E/c^2$. This result was confirmed by E. Lutz et al.[76] who showed that there is a minimum amount of heat produced per bit of erased data. This minimal heat is called the *Landauer*[77] *limit*. The equivalent mass we will call the *Landauer mass* and denote it as m_0. We will assume that a fundamental Landauer mass exists in our discussions although the precise value of the mass will not be used since we may expect all physical particle masses to be renormalized to different values when interactions are taken into account.

We will assume all fermions contain a qube within them. (As stated above bosons do not have qubes within them. We call their core a quba.) A qube is assumed to have mass m_0. The masses of fermions are modified to their known values by interactions.

It is intriguing that the mass of the electron neutrino has been measured in a variety of experiments and found to be within an order of magnitude or so larger than our estimate of the Landauer mass (as we would expect since particles acquire a 'cloud of virtual particles' due to interactions.) This 'cloud' can be expected to increase its mass above the Landauer mass. Since neutrinos only have the weak interaction it is not surprising that the increase due to interactions should not be large. The Mainz Neutrino

---

[74] In the Blaha (2018f) we called qubes iotas. However, since the name iota was previously used as a particle name many years ago it seemed reasonable to use a different name. We chose the name 'qube' for self-evident reasons. *'Qube' is pronounced 'cube.'*

[75] E. Muneyuki et al, *Nature Physics*, DOI: 10.1038/NPHYS1821.

[76] E. Lutz et al, Nature **483** (7388): 187–190,10.1038/nature10872, (2012).

[77] R. Landauer, "Irreversibility and heat generation in the computing process", IBM Journal of Research and Development **5** (3): 183–191, (1961).

Mass Experiment, for example, estimates the electron neutrino mass to be less than 2 eV. The new Karlsruhe Tritium Neutrino Experiment (September, 2019) found an upper limit of less than 1.1 eV.

A number of astronomical studies have also generated estimates of neutrino masses. In July 2010 the 3-D MegaZ DR7 galaxy survey found a limit for the combined mass of the three neutrino varieties to be less than 0.28 eV.[78] A smaller upper bound for the sum of neutrino masses, 0.23 eV, was found in March 2013 by the Planck collaboration,[79] In February 2014 a new estimate of the sum was found to be 0.320 ± 0.081 eV due to discrepancies between the Planck's measurements of the Cosmic Microwave Background, and other predictions, combined with the assumption that neutrinos are the cause of weaker gravitational lensing than implied by massless neutrinos.[80]

Thus the experimentally measured values of neutrino masses are consistent with the qube Landauer mass estimate of 0.02 eV given above. We thus assume that *a fermion particle consists of a qube with a certain mass,[81] which is renormalized, together with other features. These features will emerge later in the derivation of the complete theory.[82]*

We view Reality as ultimately a representation (or painting) of logic values evolving through interactions in time and space.[83]

9.2 Quba Cores of Fundamental Bosons

We defined a corresponding boson functional quba for each type of elementary boson. We designated a boson functional as b_s where s specifies the spin which may be 0, 1, or 2. Every boson contains a boson functional core within it. A quba has the spin of

---

[78] S. Thomas et al, "Upper Bound of 0.28 eV on Neutrino Masses from the Largest Photometric Redshift Survey", Physical Review Letters **105**: 031301 (2010).

[79] Planck Collaboration, arXiv:1303.5076 (2013).

[80] R. A. Battye et al, "Evidence for Massive Neutrinos from Cosmic Microwave Background and Lensing Observations", Phys. Rev. Lett. **112**, 051303 (2014).

[81] Leibniz first proposed the idea of logic 'particles' which he called monads. Our definition of a logic 'particle' does not include (or exclude) the presence of a spiritual part which was part of the definition of Leibniz's monads.

[82] A recent experiment claims to separate the spin part (which we identify as a logical value later) of a molecule from the rest of the molecule.

[83] Those who might suggest matter is substantial, and logic values are not, should remember that matter would be completely insubstantial if there were no forces in nature. Neutrinos which are close to insubstantial would be completely insubstantial if there were no weak interactions.

the elementary boson within which it resides. It has zero mass since bosons are typically massless prior to symmetry breaking effects.

An important consequence of the masslessness of qubas is that they have no tachyon equivalents. Note: the bare mass of qubes led to tachyons. The masslessness of qubas prevents Complex Lorentz boosts from generating tachyonic bosons.

9.3 Particle Functional Space

The functionals of elementary particles, form a point space[84] that includes all the free field fermion and boson functionals of our universe and any other universe that might exist (the Megaverse). All fundamental fermions and bosons have a corresponding particle functional. Fermion particle functionals f… are labeled with momentum k, internal symmetry quantum numbers denoted λ, and spin (See eq. 6.2 for the factorized particle functional.) Boson particle functionals b… are similarly labeled with spin s, and internal symmetry quantum numbers.

9.4 Wave Space

We assume the space-time distance between Fourier wave function expansions to be *zero* in keeping with the zero distance between particle functionals in functional space. This assumption is solidly based on the instantaneity of transformations of parts of entangled states. (No spookiness!) Separating the parts of a quantum state S into space-like separated parts S_1 and S_2.we find a change in one part causes an instantaneous change in the other part:

$$<x|S> \rightarrow <x_1|S_1><x_2||S_2> \tag{9.1}$$

irrespective of distance since the implicit functionals and Fourier expansions have no space-time separation from each other.

---

[84] Much of this chapter appears in the Blaha (2018a).

9.5 Skeleton Functional Lagrangians

If we could imagine a 'snapshot' of the universe[85] at one instant of time we could presumably enumerate all the functionals of the universe's particles. Then succeeding snapshots would show an ebb and flow of functionals as time progresses. This thought brings us to the important issue of the transformations of particle functionals in particle interactions. The simplest statement that one could make about functional transformations is that they are created and annihilated according to the interaction terms of the skeletonized Unified SuperStandard Theory (excluding quadratic terms which do not transform functionals.)

We skeletonize a lagrangian density by deleting all quadratic terms and replacing all particle fields by their corresponding functionals.[86] For example the lagrangian

$$\mathcal{L} = \bar{\psi}_C(i\gamma^\mu D_\mu - m)\psi_C(x) + b(\bar{\psi}_C\psi_C(x))^2 \qquad (9.2)$$

becomes the skeleton lagrangian

$$\mathcal{L}_S = bf^4 \qquad (9.3)$$

where f is the fermion's functional.

Thus our skeletonized lagrangian formalism describes the transitions between functionals in an interaction. This formalism is made more concrete by considering Feynman diagrams for the interactions.

9.6 Functional Interactions and Feynman Diagrams

Feynman diagrams with their in and out ordering specify the transformations between functionals more completely. A simple example shows the interaction transformations of functionals. Consider the lagrangian term

---

[85] We realize that such a snapshot is not possible since infinite velocity particles that could feed a camera this snapshot do not exist.

[86] In our construction of particle functional space we have not introduced complex conjugation of functionals for lack of a good reason. Complex conjugation takes place only in the Fourier expansion part of a quantum field. Another issue is the appearance of lagrangian terms with factors that are derivatives of fields. Since we do not do computations with skeleton lagrangians we can ignore the derivative in each such factor and simply substitute the functional. For example, $\varphi^3(\partial^\mu\varphi)^2$ becomes the quba expression b^5.

$$(\bar{\psi}\psi(x))^2(\partial^\mu\varphi)^2$$

A corresponding Feynman diagram for it is

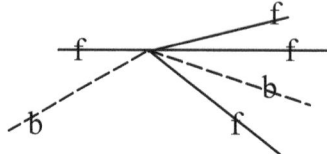

Figure 9.1.Functional Feynman diagram for the above interaction.

with qubes labeled f and qubas labeled b.

When internal symmetries are introduced then the skeletonized lagrangians and the corresponding Feynman diagram representations would be significantly more complicated.

9.7 Functional Space and Feynman Path Integrals

Functionals appear in Feynman Path Integrals and in Faddeev-Popov gauge fixing path integrals. We illustrate the use of functionals in the example:

$$Z(J) = N\int\Pi dy \;\Pi d\varphi(y)\; \exp\{i\int d^4y[\mathscr{L}(\varphi(y) + J^\mu(y)\varphi(y)]\} \qquad (9.4)$$

which in functional notation becomes

$$Z(J) = N\int\Pi d(y) \;\Pi db\; \exp\{i\int d^4y[\mathscr{L}(\varphi(y) + J^\mu(y)\varphi(y)]\} \qquad (9.5)$$

where (y) represents the Fourier expansion in the y coordinates, and with the implied inner product $\varphi(y) = (b, (y))$.

10. Monads and Observability

We have seen how particles appear to contain functionals that enable instantaneous quantum entanglement phenomena to take place without conflict with the Theory of Relativity. In this chapter we will see that observability is also directly understood within a monad framework.

Individual fundamental fermions and bosons contain functionals that we will call *monads* with certain physical properties. Fermions have qubes; bosons have qubas. We discussed their features earlier. Aggregates of fundamental particles form matter and energy. Thus matter and energy have monads within them.

10.1 Observability

We now turn to the questions of macroscopic observability and quantum observability. At both of these levels of observation monads play a decisive role. At the quantum level we have seen that monads enable instantaneous quantum phenomena to happen without conflict with Special Relativity. Quantum phenomena require observability when measured. *Thus monads, being intimately related to quantum phenomena, may also be viewed as the mechanism for observability at the quantum level.* This feature of monads was discussed in Blaha (2018f) and (2020c). Monads are then the instrument of observability at the quantum level of individual particles.

They are also the instrument of observability at the macroscopic level of aggregates of particles. Being ubiquitous monads answer questions of observability that are frequently posed, which are forms of the question:

Do events take place in the absence of observers?

Answer: YES, because the monads within particles provide a form of "default" level of observation.

10.2 Absolute Reality

Since <u>all</u> monads (functionals) exist in a space with no distance measure there is no question of disparities associated with space-time distances. Thus there is an *absolute reality* from instant to instant—all parts of the universe simultaneously exist *and are in contact with each other in principle*.[87] The universe has a unitary reality with all parts interrelated in the manner that thinkers have hypothesized for thousands of years. (For example: "The universe is one.") The events on a distant star, whose starlight we view, are as real as nearby events on earth. Aggregates of particles, which contain aggregates of monads, unfold dynamically according to physical law.

10.3 Relation to Consciousness

Consciousness has many facets. One facet is the reaction to events. In particle interactions monads transition (in general) to other monads. They do so in accord with quantum theory.

Monads thus have the general feature of reacting to events in a quantum probabilistic manner. They make "decisions" in accord with particle dynamics. Thus they have some features associated with consciousness:

1. Reaction to events

2. Selection of a path from the event point in a quantum probabilistic manner

Monads lack intelligence and decision making capacity beyond quantum dynamics. However many living creatures have similar abilities and limitations. Many creatures react to events according to a genetically determined pattern. One may characterize these creatures as partially conscious. In the next chapter we consider different levels of consciousness. The level of monad consciousness seems to be consistent with the levels of most living creatures. Mankind and other semi-intelligent species will be seen to have a higher level of consciousness.

---

[87] In particular the ubiquitous presence of gravitons—the essence of space-time—provides a universal connection mechanism as we discuss in chapter 11.

10.4 The Spark of Monads

Some thinkers (for example Leibniz) have attributed a spark of consciousness or spirit to monads This property is not determined by physical considerations and will not be considered here.

REFERENCES

Akhiezer, N. I., Frink, A. H. (tr), 1962, *The Calculus of Variations* (Blaisdell Publishing, New York, 1962).

Bjorken, J. D., Drell, S. D., 1964, *Relativistic Quantum Mechanics* (McGraw-Hill, New York, 1965).

Bjorken, J. D., Drell, S. D., 1965, *Relativistic Quantum Fields* (McGraw-Hill, New York, 1965).

Blaha, S., 1998, *Cosmos and Consciousness* (Pingree-Hill Publishing, Auburn, NH, 1998).

_____, 2002, *A Finite Unified Quantum Field Theory of the Elementary Particle Standard Model and Quantum Gravity Based on New Quantum Dimensions™ & a New Paradigm in the Calculus of Variations* (Pingree-Hill Publishing, Auburn, NH, 2002).

_____, 2003, *A Finite Unified Quantum Field Theory of the Elementary Particle Standard Model and Quantum Gravity Based on New Quantum Dimensions™ and a New Paradigm in the Calculus of Variations* (Pingree-Hill Publishing, Auburn, NH, 2003).

_____, 2004, *Quantum Big Bang Cosmology: Complex Space-time General Relativity, Quantum Coordinates™Dodecahedral Universe, Inflation, and New Spin 0, ½, 1 & 2 Tachyons & Imagyons* (Pingree-Hill Publishing, Auburn, NH, 2004).

_____, 2005a, *Quantum Theory of the Third Kind: A New Type of Divergence-free Quantum Field Theory Supporting a Unified Standard Model of Elementary Particles and Quantum Gravity based on a New Method in the Calculus of Variations* (Pingree-Hill Publishing, Auburn, NH, 2005).

_____, 2005b, *The Metatheory of Physics Theories, and the Theory of Everything as a Quantum Computer Language* (Pingree-Hill Publishing, Auburn, NH, 2005).

_____, 2005c, *The Equivalence of Elementary Particle Theories and Computer Languages: Quantum Computers, Turing Machines, Standard Model, Superstring Theory, and a Proof that Gödel's Theorem Implies Nature Must Be Quantum* (Pingree-Hill Publishing, Auburn, NH, 2005).

_____, 2006a, *The Foundation of the Forces of Nature* (Pingree-Hill Publishing, Auburn, NH, 2006).

_____, 2006b, *A Derivation of ElectroWeak Theory based on an Extension of Special Relativity; Black Hole Tachyons; & Tachyons of Any Spin.* (Pingree-Hill Publishing, Auburn, NH, 2006).

_____, 2007a, *Physics Beyond the Light Barrier: The Source of Parity Violation, Tachyons, and A Derivation of Standard Model Features* (Pingree-Hill Publishing, Auburn, NH, 2007).

_____, 2007b, *The Origin of the Standard Model: The Genesis of Four Quark and Lepton Species, Parity Violation, the ElectroWeak Sector, Color SU(3), Three Visible Generations of Fermions, and One Generation of Dark Matter with Dark Energy* (Pingree-Hill Publishing, Auburn, NH, 2007).

_____, 2008a, *A Direct Derivation of the Form of the Standard Model From GL(16) (Pingree-Hill Publishing, Auburn, NH, 2008).*

_____, 2008b, *A Complete Derivation of the Form of the Standard Model With a New Method to Generate Particle Masses Second Edition* (Pingree-Hill Publishing, Auburn, NH, 2008)

_____, 2009, *The Algebra of Thought & Reality: The Mathematical Basis for Plato's Theory of Ideas, and Reality Extended to Include A Priori Observers and Space-Time Second Edition* (Pingree-Hill Publishing, Auburn, NH, 2009).

_____, 2010a, *Operator Metaphysics: A New Metaphysics Based on a New Operator Logic and a New Quantum Operator Logic that Lead to a Mathematical Basis for Plato's Theory of Ideas and Reality* (Pingree-Hill Publishing, Auburn, NH, 2010).

_____, 2010b, *The Standard Model's Form Derived from Operator Logic, Superluminal Transformations and GL(16)* (Pingree-Hill Publishing, Auburn, NH, 2010).

_____, 2010c, *SuperCivilizations: Civilizations as Superorganisms* (McMann-Fisher Publishing, Auburn, NH, 2010).

_____, 2011a, *21st Century Natural Philosophy Of Ultimate Physical Reality* (McMann-Fisher Publishing, Auburn, NH, 2011).

_____, 2011b, *All the Universe! Faster Than Light Tachyon Quark Starships & Particle Accelerators with the LHC as a Prototype Starship Drive Scientific Edition* (Pingree-Hill Publishing, Auburn, NH, 2011).

_____, 2011c, *From Asynchronous Logic to The Standard Model to Superflight to the Stars* (Blaha Research, Auburn, NH, 2011).

_____, 2012a, *From Asynchronous Logic to The Standard Model to Superflight to the Stars volume 2: Superluminal CP and CPT, U(4) Complex General Relativity and The Standard Model, Complex Vierbein General Relativity, Kinetic Theory, Thermodynamics* (Blaha Research, Auburn, NH, 2012).

_____, 2012b, *Standard Model Symmetries, And Four And Sixteen Dimension Complex Relativity; The Origin Of Higgs Mass Terms* (Blaha Reasearch, Auburn, NH, 2012).

_____, 2013a, *Multi-Stage Space Guns, Micro-Pulse Nuclear Rockets, and Faster-Than-Light Quark-Gluon Ion Drive Starships* (Blaha Research, Auburn, NH, 2013).

_____, 2013b, *The Bridge to Dark Matter; A New Sister Universe; Dark Energy; Inflatons; Quantum Big Bang; Superluminal Physics; An Extended Standard Model Based on Geometry* (Blaha Reasearch, Auburn, NH, 2013).

_____, 2014a, *Universes and Megaverses: From a New Standard Model to a Physical Megaverse; The Big Bang; Our Sister Universe's Wormhole; Origin of the Cosmological Constant, Spatial Asymmetry of the Universe, and its Web of Galaxies; A Baryonic Field between Universes and Particles; Megaverse Extended Wheeler-DeWitt Equation* (Blaha Reasearch, Auburn, NH, 2014).

_____, 2014b, *All the Megaverse! Starships Exploring the Endless Universes of the Cosmos Using the Baryonic Force* (Blaha Research, Auburn, NH, 2014).

_____, 2014c, *All the Megaverse! II Between Megaverse Universes: Quantum Entanglement Explained by the Megaverse Coherent Baryonic Radiation Devices – PHASERs Neutron Star Megaverse Slingshot Dynamics Spiritual and UFO Events, and the Megaverse Microscopic Entry into the Megaverse* (Blaha Research, Auburn, NH, 2014).

_____, 2015a, *PHYSICS IS LOGIC PAINTED ON THE VOID: Origin of Bare Masses and The Standard Model in Logic, U(4) Origin of the Generations, Normal and Dark Baryonic Forces, Dark Matter, Dark Energy, The Big Bang, Complex General Relativity, A Megaverse of Universe Particles* (Blaha Research, Auburn, NH, 2015).

_____, 2015b, *PHYSICS IS LOGIC Part II: The Theory of Everything, The Megaverse Theory of Everything, U(4)⊗U(4) Grand Unified Theory (GUT), Inertial Mass = Gravitational Mass, Unified Extended Standard Model and a New Complex General Relativity with Higgs Particles, Generation Group Higgs Particles* (Blaha Research, Auburn, NH, 2015).

_____, 2015c, *The Origin of Higgs ("God") Particles and the Higgs Mechanism: Physics is Logic III, Beyond Higgs – A Revamped Theory With a Local Arrow of Time, The Theory of Everything Enhanced, Why Inertial Frames are Special, Universes of the Mind* (Blaha Research, Auburn, NH, 2015).

_____, 2015d, *The Origin of the Eight Coupling Constants of The Theory of Everything: U(8) Grand Unified Theory of Everything (GUTE), S^8 Coupling Constant Symmetry, Space-Time Dependent Coupling Constants, Big Bang Vacuum Coupling Constants, Physics is Logic IV* (Blaha Research, Auburn, NH, 2015).

_____, 2016a, *New Types of Dark Matter, Big Bang Equipartition, and A New U(4) Symmetry in the Theory of Everything: Equipartition Principle for Fermions, Matter is 83.33% Dark,*

Penetrating the Veil of the Big Bang, Explicit QFT Quark Confinement and Charmonium, Physics is Logic V (Blaha Research, Auburn, NH, 2016).

_____, 2016b, *The Periodic Table of the 192 Quarks and Leptons in The Theory of Everything: The U(4) Layer Group, Physics is Logic VI* (Blaha Research, Auburn, NH, 2016).

_____, 2016c, *New Boson Quantum Field Theory, Dark Matter Dynamics, Dark Matter Fermion Layer Mixing, Genesis of Higgs Particles, New Layer Higgs Masses, Higgs Coupling Constants, Non-Abelian Higgs Gauge Fields, Physics is Logic VII* (Blaha Research, Auburn, NH, 2016).

_____, 2016d, *Unification of the Strong Interactions and Gravitation: Quark Confinement Linked to Modified Short-Distance Gravity; Physics is Logic VIII* (Blaha Research, Auburn, NH, 2016).

_____, 2016e, *MoND: Unification of the Strong Interactions and Gravitation II, Quark Confinement Linked to Large-Scale Gravity, Physics is Logic IX* (Blaha Research, Auburn, NH, 2016).

_____, 2016f, *CQ Mechanics: A Unification of Quantum & Classical Mechanics, Quantum/Semi-Classical Entanglement, Quantum/Classical Path Integrals, Quantum/Classical Chaos* (Blaha Research, Auburn, NH, 2016).

_____, 2016g, *GEMS: Unified Gravity, ElectroMagnetic and Strong Interactions: Manifest Quark Confinement, A Solution for the Proton Spin Puzzle, Modified Gravity on the Galactic Scale* (Pingree Hill Publishing, Auburn, NH, 2016).

_____, 2016h, *Unification of the Seven Boson Interactions based on the Riemann-Christoffel Curvature Tensor* (Pingree Hill Publishing, Auburn, NH, 2016).

_____, 2017a, *Unification of the Eleven Boson Interactions based on 'Rotations of Interactions'* (Pingree Hill Publishing, Auburn, NH, 2017).

_____, 2017b, *The Origin of Fermions and Bosons, and Their Unification* (Pingree Hill Publishing, Auburn, NH, 2017).

_____, 2017c, *Megaverse: The Universe of Universes* (Pingree Hill Publishing, Auburn, NH, 2017).

_____, 2017d, *SuperSymmetry and the Unified SuperStandard Model* (Pingree Hill Publishing, Auburn, NH, 2017).

_____, 2017e, *From Qubits to the Unified SuperStandard Model with Embedded SuperStrings: A Derivation* (Pingree Hill Publishing, Auburn, NH, 2017).

_____, 2017f, *The Unified SuperStandard Model in Our Universe and the Megaverse: Quarks, ... ,* (Pingree Hill Publishing, Auburn, NH, 2017).

_____, 2018a, *The Unified SuperStandard Model and the Megaverse SECOND EDITION A Deeper Theory based on a New Particle Functional Space that Explicates Quantum Entanglement Spookiness (Volume 1)* (Pingree Hill Publishing, Auburn, NH, 2018).

_____, 2018b, *Cosmos Creation: The Unified SuperStandard Model, Volume 2, SECOND EDITION* (Pingree Hill Publishing, Auburn, NH, 2018).

_____, 2018c, *God Theory (*Pingree Hill Publishing, Auburn, NH, 2018).

_____, 2018d, *Immortal Eye: God Theory: Second Edition* (Pingree Hill Publishing, Auburn, NH, 2018).

_____, 2018e, *Unification of God Theory and Unified SuperStandard Model THIRD EDITION* (Pingree Hill Publishing, Auburn, NH, 2018).

_____, 2019a, *Calculation of: QED α = 1/137, and Other Coupling Constants of the Unified SuperStandard Theory* (Pingree Hill Publishing, Auburn, NH, 2019).

_____, 2019b, *Coupling Constants of the Unified SuperStandard Theory SECOND EDITION* (Pingree Hill Publishing, Auburn, NH, 2019).

_____, 2019c, *New Hybrid Quantum Big_Bang–Megaverse_Driven Universe with a Finite Big Bang and an Increasing Hubble Constant* (Pingree Hill Publishing, Auburn, NH, 2019).

_____, 2019d, *The Universe, The Electron and The Vacuum* (Pingree Hill Publishing, Auburn, NH, 2019).

_____, 2019e, *Quantum Big Bang – Quantum Vacuum Universes (Particles)* (Pingree Hill Publishing, Auburn, NH, 2019).

_____, 2019f, *The Exact QED Calculation of the Fine Structure Constant Implies ALL 4D Universes have the Same Physics/Life Prospects* (Pingree Hill Publishing, Auburn, NH, 2019).

_____, 2019g, *Unified SuperStandard Theory and the SuperUniverse Model: The Foundation of Science* (Pingree Hill Publishing, Auburn, NH, 2019).

_____, 2020a, *Quaternion Unified SuperStandard Theory (The QUeST) and Megaverse Octonion SuperStandard Theory (MOST)* (Pingree Hill Publishing, Auburn, NH, 2020).

_____, 2020b, *United Universes Quaternion Universe - Octonion Megaverse* (Pingree Hill Publishing, Auburn, NH, 2020).

_____, 2020c, *Unified SuperStandard Theories for Quaternion Universes & The Octonion Megaverse* (Pingree Hill Publishing, Auburn, NH, 2020).

_____, 2020d, *The Essence of Eternity: Quaternion & Octonion SuperStandard Theories* (Pingree Hill Publishing, Auburn, NH, 2020).

_____, 2020e, *The Essence of Eternity II* (Pingree Hill Publishing, Auburn, NH, 2020).

Eddington, A. S., 1952, *The Mathematical Theory of Relativity* (Cambridge University Press, Cambridge, U.K., 1952).

Fant, Karl M., 2005, *Logically Determined Design: Clockless System Design With NULL Convention Logic* (John Wiley and Sons, Hoboken, NJ, 2005).

Feinberg, G. and Shapiro, R., 1980, *Life Beyond Earth: The Intelligent Earthlings Guide to Life in the Universe* (William Morrow and Company, New York, 1980).

Gelfand, I. M., Fomin, S. V., Silverman, R. A. (tr), 2000, *Calculus of Variations* (Dover Publications, Mineola, NY, 2000).

Giaquinta, M., Modica, G., Souchek, J., 1998, *Cartesian Coordinates in the Calculus of Variations* Volumes I and II (Springer-Verlag, New York, 1998).

Giaquinta, M., Hildebrandt, S., 1996, *Calculus of Variations* Volumes I and II (Springer-Verlag, New York, 1996).

Gradshteyn, I. S. and Ryzhik, I. M., 1965, *Table of Integrals, Series, and Products* (Academic Press, New York, 1965).

Heitler, W., 1954, *The Quantum Theory of Radiation* (Claendon Press, Oxford, UK, 1954).

Huang, Kerson, 1992, *Quarks, Leptons & Gauge Fields 2nd Edition* (World Scientific Publishing Company, Singapore, 1992).

Jost, J., Li-Jost, X., 1998, *Calculus of Variations* (Cambridge University Press, New York, 1998).

Kaku, Michio, 1993, *Quantum Field Theory*, (Oxford University Press, New York, 1993).

Kirk, G. S. and Raven, J. E., 1962, *The Presocratic Philosophers* (Cambridge University Press, New York, 1962).

Landau, L. D. and Lifshitz, E. M., 1987, *Fluid Mechanics 2nd Edition*, (Pergamon Press, Elmsford, NY, 1987).

Misner, C. W., Thorne, K. S., and Wheeler, J. A., 1973, *Gravitation* (W. H. Freeman, New York, 1973).

Rescher, N., 1967, *The Philosophy of Leibniz* (Prentice-Hall, Englewood Cliffs, NJ, 1967).

Rieffel, Eleanor and Polak, Wolfgang, 2014, *Quantum Computing* (MIT Press, Cambridge, MA, 2014).

Riesz, Frigyes and Sz.-Nagy, Béla, 1990, *Functional Analysis* (Dover Publications, New York, 1990).

Sagan, H., 1993, *Introduction to the Calculus of Variations* (Dover Publications, Mineola, NY, 1993).

Sakurai, J. J., 1964, *Invariance Principles and Elementary Particles* (Princeton University Press, Princeton, NJ, 1964).

Sorokin, Pitirim, 1941, *Social and Cultural Dynamics* (Porter Sargent Publishers, Boston, MA, 1941).

Streater, R. F. and Wightman, A. S., 2000, *PCT, Spin, Statistics, and All That* (Princeton University Press, Princeton, NJ 2000).

Weinberg, S., 1972, *Gravitation and Cosmology* (John Wiley and Sons, New York, 1972).

Weinberg, S., 1995, *The Quantum Theory of Fields Volume I* (Cambridge University Press, New York, 1995).

Weinberg, S., 2000, *The Quantum Theory of Fields Volume III Supersymmetry* (Cambridge University Press, New York, 2000).

Weyl, H., 1950, *Space, Time, Matter* (Dover, New York, 1950).

Weyl, H., (Tr. S. Pollard et al), 1987, *The Continuum* (Dover Publications, New York, 1987).

INDEX

162

About the Author

Stephen Blaha is a well-known Physicist and Man of Letters with interests in Science, Society and civilization, the Arts, and Technology. He had an Alfred P. Sloan Foundation scholarship in college. He received his Ph.D. in Physics from Rockefeller University. He has served on the faculties of several major universities. He was also a Member of the Technical Staff at Bell Laboratories, a manager at the Boston Globe Newspaper, a Director at Wang Laboratories, and President of Blaha Software Inc. and of Janus Associates Inc. (NH).

Among other achievements he was a co-discoverer of the "r potential" for heavy quark binding developing the first (and still the only demonstrable) non-Aeolian gauge theory with an "r" potential; first suggested the existence of topological structures in superfluid He-3; first proposed Yang-Mills theories would appear in condensed matter phenomena with non-scalar order parameters; first developed a grammar-based formalism for quantum computers and applied it to elementary particle theories; first developed a new form of quantum field theory without divergences (thus solving a major 60 year old problem that enabled a unified theory of the Standard Model and Quantum Gravity without divergences to be developed); first developed a formulation of complex General Relativity based on analytic continuation from real space-time; first developed a generalized non-homogeneous Robertson-Walker metric that enabled a quantum theory of the Big Bang to be developed without singularities at t = 0; first generalized Cauchy's theorem and Gauss' theorem to complex, curved multi-dimensional spaces; received Honorable Mention in the Gravity Research Foundation Essay Competition in 1978; first developed a physically acceptable theory of faster-than-light particles; first derived a composition of extremums method in the Calculus of Variations; first quantitatively suggested that inflationary periods in the history of the universe were not needed; first proved Gödel's Theorem implies Nature must be quantum; provided a new alternative to the Higgs Mechanism, and Higgs particles, to generate masses; first showed how to resolve logical paradoxes including Gödel's Undecidability Theorem by developing Operator Logic and Quantum Operator Logic; first developed a quantitative harmonic oscillator-like model of the life cycle, and interactions, of civilizations; first showed how equations describing superorganisms also apply to civilizations. A recent book shows his theory applies successfully to the past 14 years of history and to *new* archaeological data on Andean and Mayan civilizations as well as Early Anatolian and Egyptian civilizations.

He first developed an axiomatic derivation of the form of The Standard Model from geometry – space-time properties – The Unified SuperStandard Model. It unifies all the known forces of Nature. It also has a Dark Matter sector that includes a Dark ElectroWeak sector with Dark doublets and Dark gauge interactions. It uses quantum coordinates to remove infinities that crop up in most interacting quantum field theories and additionally to remove the infinities that appear in the Big Bang and generate inflationary growth of the universe. It shows

gravity has a MOND-like form without sacrificing Newton's Laws. It relates the interactions of the MOND-like sector of gravity with the r-potential of Quark Confinement. The axioms of the theory lead to the question of their origin. We suggest in the preceding edition of this book it can be attributed to an entity with God-like properties. We explore these properties in "God Theory" and show they predict that the Cosmos exists forever although individual universes (or incarnations of our universe) "come and go." Several other important results emerge from God Theory such a functionally triune God. The Unified SuperStandard Theory has many other important parts described in the Current Edition of *The Unified SuperStandard Theory* and expanded in subsequent volumes.

Blaha has had a major impact on a succession of elementary particle theories: his Ph.D. thesis (1970), and papers, showed that quantum field theory calculations to all orders in ladder approximations could not give scaling deep inelastic electron-nucleon scattering. He later showed the eigenvalue equation for the fine structure constant α in Johnson-Baker-Willey QED had a zero at $\alpha = 1$ not 1/137 by solving the Schwinger-Dyson equations to all orders in an approximation that agreed with exact results to 4th order in α thus ending interest in this theory. In 1979 at Prof. Ken Johnson's (MIT) suggestion he calculated the proton-neutron mass difference in the MIT bag model and found the result had the wrong sign reducing interest in the bag model. These results all appear in Physical Review papers. In the 2000's he repeatedly pointed out the shortcomings of SuperString theory and showed that The Standard Model's form could be derived from space-time geometry by an extension of Lorentz transformations to faster than light transformations. This deeper space-time basis greatly increases the possibility that it is part of THE fundamental theory. Recently, Blaha showed that the Weak interactions differed significantly from the Strong, electromagnetic and gravitation interactions in important respects while these interactions had similar features, and suggested that ElectroWeak theory, which is essentially a glued union of the Weak interactions and Electromagnetism, possibly modulo unknown Higgs particle features, be replaced by a unified theory of the other interactions combined with a stand-alone Weak interaction theory. Blaha also showed that, if Charmonium calculations are taken seriously, the Strong interaction coupling constant is only a factor of five larger than the electromagnetic coupling constant, and thus Strong interaction perturbation theory would make sense and yield physically meaningful results.

In graduate school (1965-71) he wrote substantial papers in elementary particles and group theory: The Inelastic E- P Structure Functions in a Gluon Model. Phys. Lett. B40:501-502,1972; Deep-Inelastic E-P Structure Functions In A Ladder Model With Spin 1/2 Nucleons, Phys.Rev. D3:510-523,1971; Continuum Contributions To The Pion Radius, Phys. Rev. 178:2167-2169,1969; Character Analysis of U(N) and SU(N), J. Math. Phys. 10, 2156 (1969); and The Calculation of the Irreducible Characters of the Symmetric Group in Terms of the Compound Characters, (Published as Blaha's Lemma in D. E. Knuth's book: *The Art of Computer Programming Vols. 1 – 4*).

In the early 1980's Blaha was also a pioneer in the development of UNIX for financial, scientific and Internet applications: benchmarked UNIX versions showing that block size was critical for UNIX performance, developing financial modeling software, starting database benchmarking comparison studies, developing Internet-like UNIX networking (1982) and developing a hybrid shell programming technique (1982) that was a precursor to the PERL programming language. He was also the manager of the AT&T ten-year future products development database. His work helped lead to commercial UNIX on computers such as Sun Micros, IBM AIX minis, and Apple computers.

In the 1980's he pioneered the development of PC Desktop Publishing on laser printers and was nominated for three "Awards for Technical Excellence" in 1987 by PC Magazine for PC software products that he designed and developed.

Recently he has developed a theory of Megaverses – actual universes of which our universe is one – with quantum particle-like properties based on the Wheeler-DeWitt equation of Quantum Gravity. He has developed a theory of a baryonic force, which had been conjectured many years ago, and estimated the strength of the force based on discrepancies in measurements of the gravitational constant G. This force, operative in D-dimensional space, can be used to escape from our universe in "uniships" which are the equivalent of the faster-than-light starships proposed in the author's earlier books. Thus travel to other universes, as well as to other stars is possible.

Blaha also considered the complexified Wheeler-DeWitt equation and showed that its limitation to real-valued coordinates and metrics generated a Cosmological Constant in the Einstein equations.

The author has also recently written a series of books on the serious problems of the United States and their solution as well as a book on the decline of Mankind that will follow from current social and genetic trends in Mankind.

In the past twenty years Dr. Blaha has written over 80 books on a wide range of topics. Some recent major works are: *From Asynchronous Logic to The Standard Model to Superflight to the Stars*, *All the Universe!*, *SuperCivilizations: Civilizations as Superorganisms*, *America's Future: an Islamic Surge, ISIS, al Qaeda, World Epidemics, Ukraine, Russia-China Pact, US Leadership Crisis*, *The Rises and Falls of Man – Destiny – 3000 AD: New Support for a Superorganism MACRO-THEORY of CIVILIZATIONS From CURRENT WORLD TRENDS and NEW Peruvian, Pre-Mayan, Mayan, Anatolian, and Early Egyptian Data, with a Projection to 3000 AD*, and *Mankind in Decline: Genetic Disasters, Human-Animal Hybrids, Overpopulation, Pollution, Global Warming, Food and Water Shortages, Desertification, Poverty, Rising Violence, Genocide, Epidemics, Wars, Leadership Failure*.

He has taught approximately 4,000 students in undergraduate, graduate, and postgraduate corporate education courses primarily in major universities, and large companies and government agencies.

Recently he developed a quantum theory, The Unified SuperStandard Theory (UST), which describes elementary particles in detail without the difficulties of conventional quantum field theory. He found that the internal symmetries of this theory could be exactly derived from a 32 dimension complex quaternion theory called QUeST. He further found that a 32 dimension complex octonion theory (MOST) describes the Megaverse. It can hold QUeST universes such as our own universe. It has an internal symmetry structure which is a superset of the QUeST internal symmetries.

Lightning Source UK Ltd.
Milton Keynes UK
UKHW031827030622
403946UK00004B/171